Detecting Ophthalmic Nerve Damage with Smart Analytic Mechanisms

Chika

Detecting Ophthalmic Nerve Damage With Smart Analytic Mechanisms

First Edition June 2025

Written by Chika

TABLE OF CONTENTS

CHAPTER 1

INTRODUCTION

1.1 OVERVIEW OF GLAUCOMA DIAGNOSIS

An eye, the organ of a visual system, responds to light and provides vision. The optic nerve is a part of the eye that is recognized to be a unique area of sight. It is effective in conveying object representation from the eye to the brain. Glaucoma is one of the most common eye disorders that is widespread among the people in recent years. Intraocular pressure, or increased pressure inside the eye, damages the optic nerve. When the optic nerve is injured, vision loss can occur and blood flow may be disrupted. It results in eye disorder like glaucoma. Figure 1.1 represents a normal human eye and that with glaucoma (Akbar *et al.* 2022).

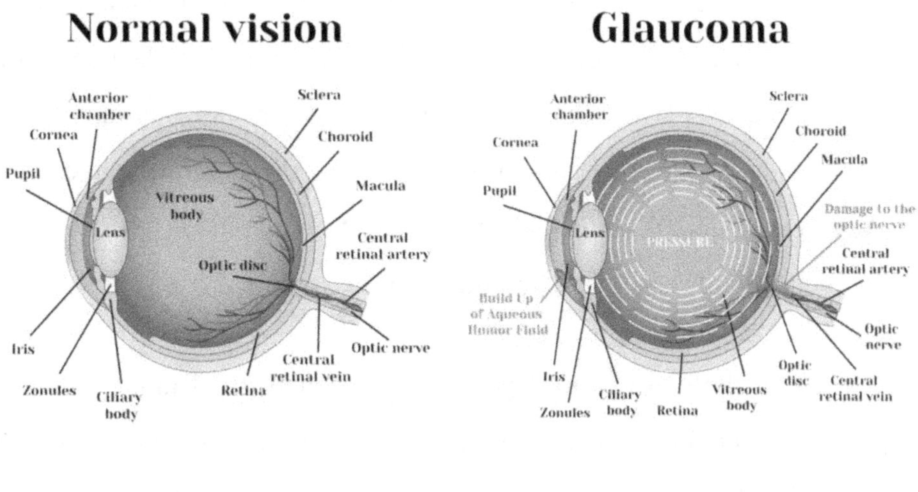

Figure 1.1 Images of a normal and a glaucoma eye

Thus, glaucoma is a word that refers to an eye disorder characterized by progressive optic nerve damage with a huge rise in Intra-Ocular Pressure (IOP) (Jonas *et al.* 2017). To keep eyes moist and healthy, eyes generate a fluid known as aqueous humor. Whenever the fluid is released, an equal amount of fluid should be drained away, ensuring a balance between fluid production and outflow. The pressure inside the eye rises when there is an imbalance caused by either overproduction or incorrect drainage of the fluid. The optic nerve is damaged by this pressure, which is termed as IOP and as a result, glaucoma develops. Thus, increased pressure damages the optic nerve when fluid accumulates as a result of poor drainage or surplus of fluid. Glaucoma and diabetic retinopathy are the two most common causes of blindness in the world. Researchers (Allison *et al.* 2020) states that there is a gradual growth of glaucoma disease among people. Hence in the day-to-day lives of human beings, early identification of the disease is vital. Glaucoma is one of the most common causes of blindness in persons over the age of 60 (Malik *et al.* 2020). It can affect anyone at any age, but it is more common in the elderly people. It is important for human beings to inspect their eyes on a regular basis in order to protect vision.

Glaucoma can be broadly distinguished into open-angle glaucoma and angle-closure glaucoma. One of the most common glaucoma is the open angle glaucoma which has no symptoms and is more common worldwide. This kind of glaucoma is observed among 90% of all glaucoma patients (Veena *et al.* 2020). Likewise, angle-closure glaucoma is a serious eye disease that necessitates early medical intervention.

There are various possible consequences of glaucoma such as eye pain, hypertension, headache, redness in eye, inflammation of eye and blurred vision. Medications and surgery can be used to treat glaucoma. Usually five typical glaucoma tests namely tonometry, ophthalmoscopy, perimetry, gonioscopy and pachymetry are included in a regular checkup. During the diagnosis of glaucoma, the inner eye pressure is measured with a tonometry test, while the shape and color of the optic nerve is evaluated with an ophthalmoscopy. Perimetry, a visual field test, is used to evaluate the entire field of vision, gonioscopy is used to investigate the angle in the eye and pachymetry is utilized to assess the thickness of the cornea.

Glaucoma can also be diagnosed using the clinical parameters such as Cup- to-Disc Ratio (CDR) (Jonas *et al.* 2000), Rim to Disc Area ratio (RDAR) (Jonas *et al.* 1998) and disc diameter (Hancox 1999). Out of these three metrics, CDR is the most widely accepted and used one by clinical specialists. A higher CDR indicates a higher risk of glaucoma while a lower CDR indicates a decreased risk of glaucoma. For measuring CDR, accurate optic disc and optic cup segmentations are required. Some approaches use 3-D OCT (Optical Coherence Tomography) to automatically measure the disc and cup (Lee *et al.* 2010). However, OCT is not widely accessible due to its expensive cost; many doctors still refer to the fundus image. A number of methods have been proposed for segmenting the optic disc and optic cup from a fundus image (Joshi *et al.* 2011). Boundary detection, region segmentation and color & contrast thresholding techniques are the most common segmentation techniques (Almazroa *et al.* 2015).

As manual diagnosis by a well-trained clinic specialist takes time and involves much expense, it is not suitable for large-scale population screening. As a result, automatic assessment methods are developed, in which computers are trained to classify the images into the respective category. Deep learning approaches have recently been shown to provide highly discriminative representations, which have helped with a variety of computer vision problems. It is gaining popularity in areas such as medical image processing, speech recognition, pattern recognition and natural language processing. With the advancement in technologies, computers have been educated to do certain tasks by processing vast amounts of data like images, text, sound and so on.

Deep Neural Networks being having a lot of success since 2012. Deep learning is used to improve the accuracy of many real-world problems. It identifies the features of the input on its own. At present, deep learning methods have gained much attention in the medical field for various applications like biomarker segmentation, lesion segmentation, image synthesis and disease diagnosis. Hence deep learning has been employed to diagnose glaucoma.

1.2 STRUCTURE OF A HUMAN EYE

The eye is one of the five senses, connecting the human beings to the rest of the body. Unlike the other organs, the eye is a light-sensitive organ. The eye usually measures approximately 1 inch wide (2.54 cm) and 0.9 inch tall (2.3 cm). A cross section of the human eye is shown in Figure 1.2 (Shirzadfar & Amirzadeh 2019) along with its major components. The sclera is a visible white component that protects the inner organs. Likewise, the conjunctiva protects the sclera and it lubricates the eyes. It keeps the eyes moist by secreting fluid and tears.

Cornea is the front component of the eye that surrounds the pupil and iris. It refracts the light with the lens. The primary role of the iris is to regulate the size of the pupil in accordance with the source of light. The pupil is the little opening at the center of the iris. It lets light in and focuses it on the retina. Between both the cornea and lens, there is a fluid called aqueous humour that nourishes the eye. Further, vitreous humour is a jelly-like translucent material found between the lens and the retina. In between the retina and the sclera, the choroid is a circulatory layer of the eye which encompasses the iris and the lens-attached ciliary body. It includes the blood vessels which provide blood to the eye's components.

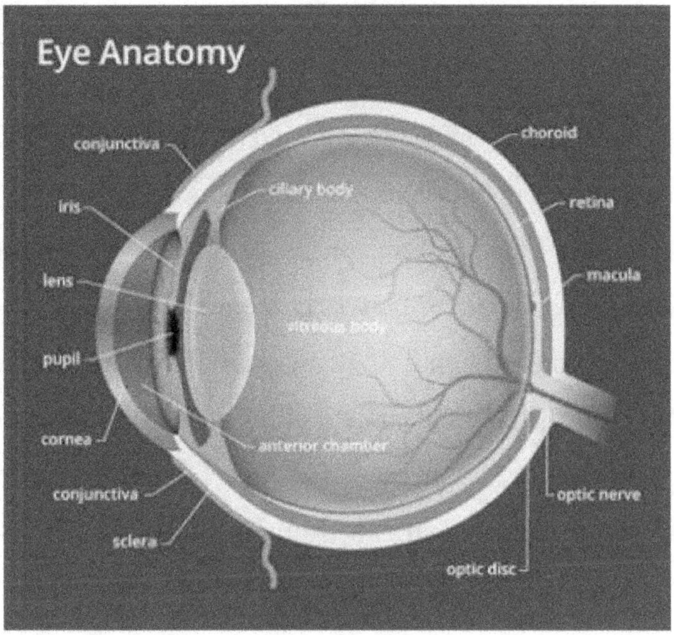

Figure 1.2 Anatomy of a Human Eye

Similarly, the retina is a thin layer of a photosensitive tissue made up of several layers of cells designed to absorb light and to convert it into neural signals. It is approximately 0.25mm thick. The macula, in which a detailed central vision is formed, is a highly light-sensitive area of the retina located at the central region measuring 5 to 6 mm diameter. After travelling through the cornea, pupil and lens, the reflected light from an object is concentrated on the retina in the same way that the light passes through the lens or film of a camera. The optic nerve is found at the back of the eyes. All the nerve signals from the retina to the brain are transmitted via the optic nerves, which are responsible for perception. The optic disc is also known as the Optic Nerve Head (ONH). It is a round/oval 'plug-hole' through which over a million nerve fibers pass through the lamina cribrosa, a sieve-like sheet. These fibers gather behind the eye to form the optic nerve, which travels to the brain. The normal CDR should be under 0.5. If the value is around 0.6, then it will be considered as having mild glaucoma and if it is around 1.01 (Mohamed *et al.* 2020), then it will be considered as a severe glaucoma case. The normal ocular pressure ranges from 12 to 22 mm Hg.

Glaucoma is an intensified nerve disorder that occurs due to the enlarged optic disc cupping and notching of neuroretinal rims. From several reports, it has been found that one of the leading causes of blindness in human beings is due to glaucoma. So, an early identification of optic disc changes is needed for glaucoma diagnosis. Screening of fundus images is extremely important in the medical field since it allows for early detection of eye related diseases. Many features can be found in a fundus image to help diagnose glaucoma. CDR, Neuroretinal rim, Retinal thickness and Inferior, Superior, Nasal and Temporal (ISNT) rule are examples of these features that are used to detect eye related diseases. The machine learning and deep learning algorithms that can evaluate any of these features and detect glaucoma from a fundus image would be a useful one for detecting glaucoma early or for controlling its progression.

1.3 VISUAL SENSE AND NEURAL ASPECTS

The optic nerve, which enters the eyeball and links the rods and cones at the rear of the eye, is responsible for human vision. Dendrites (inputs) and lengthy axons with branched terminals (outputs) make up neurons. During neuronal communication, synapses are used to transfer impulses. The axons of the ganglion cells inside the eye make up the optic nerve. They transmit the visual data to the brain. The pathway from the eyes to the brain is depicted in Figure 1.3 (Armstrong & Cubbidge 2014). Further, the fibers from each eye's temporal halves pass on to the brain on the same side while that from each eye's nasal halves cross over at the optic chiasm (L- left R- right).

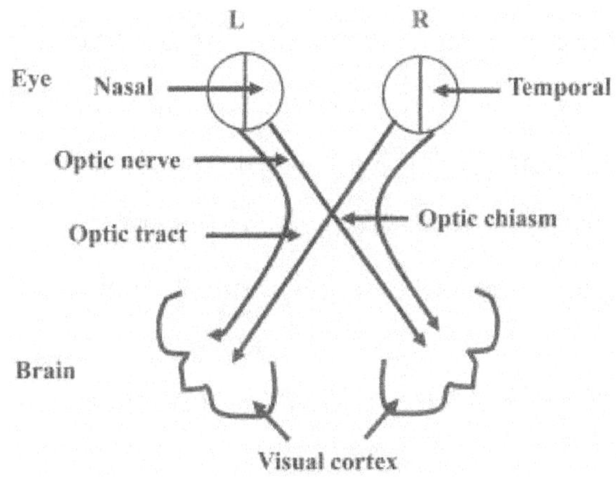

Figure 1.3 Neural pathway for vision

Transmission of signals depends on the diffusion of chemicals across interfaces and these chemicals stimulate or inhibit the receiving neurons. Initially, the optic nerves bundle the axons of the ganglion cells on one side of the retina. In the meanwhile, bipolar cells connect the rods and cones to the ganglion cells. Horizontal cells connect the rods and cones. Moreover, several ganglion cells are linked by amacrine cells. The horizontal cells in the retina are responsible for combining signals from the neighbouring receptors as well as creating a receptive field with opposing reactions at the centre and the periphery. The difference in illumination between the centers and the peripheries in a non-uniform illumination creates stimulations. In the lateral geniculate bodies and in the visual cortex, there are further groups of receptive field responses associated with detecting directional edges.

Rods and cones send optical impulses to the brain via the fibers present in the optic nerves. The optic chiasm is a location where the optic nerves from each eye meet slightly below the brain. The optic nerves are crossed at the optic chiasm. The information from the right visual field (which comes from both the eyes) is delivered to the left side of the brain while the information from the left visual field is sent to the right side of the brain, at the location of the optic chiasm. The optical nerves in the brain end halfway back at the lateral geniculate bodies, from which visual impulses are transmitted to the visual cortex. The visual cortex is only the initial step of perception, where data become accessible and it still has an architecture that is comparable to the retina. A corpus callosum connects the halves of the visual field and links the two cerebral hemispheres of visual regions.

1.4 DISEASES AND COMMON DISORDERS OF THE EYE

Eye illnesses are diseases or disorders that cause the destruction of an eye tissue or any other part of the eye. There are different eye disorders, some of which are mild and may not endure for extended periods of time while the others might result in permanent vision loss. Age related macular degeneration, cataracts, diabetic retinopathy and glaucoma are the leading causes of blindness and impaired vision. Strabismus and amblyopia are the two most frequent eye disorders.

1.4.1 Age related Macular Degeneration (AMD)

The Macula is responsible for acute vision in everyday life. Central vision is required to view objects clearly and execute routine tasks like driving and reading. Macular degeneration is a condition that affects the macula, which is the central portion of the retina. Small patches in the central vision caused by age related macular degeneration eventually lead to blindness. As it happens with ageing, macular deterioration is sometimes referred to as Age related Macular Degeneration (AMD). Age, high blood pressure, obesity and sun exposure are all factors that contribute to macular degeneration. Blurred vision and black or empty patch at the centre of vision are a few symptoms of macular degeneration. Early AMD have a few tiny drusen in the retina. Figure 1.4 shows the fundus image with mascular drusen (Shahsuvaryan *et al.* 2005).

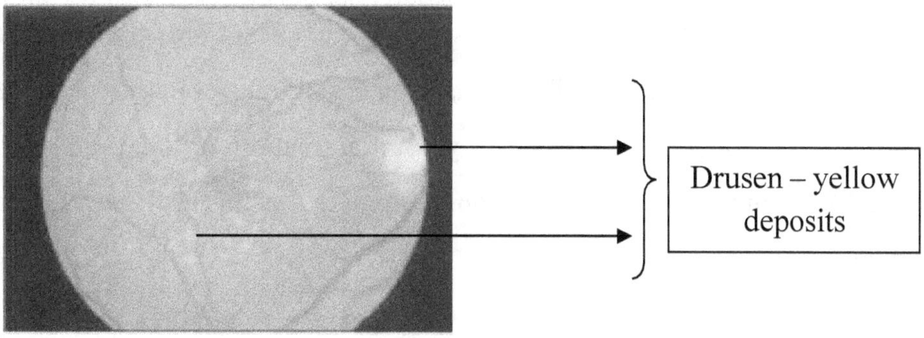

Figure 1.4 Macular drusen

Dry macular degeneration and wet macular degeneration are two different types of macular degeneration. Wet AMD occurs when the aberrant blood vessels beneath the retina begin to form beneath the macula, resulting in blood and fluid leaks. Continuous leakage, bleeding and thickening of these blood vessels induce harm and lead to central vision loss quickly. As a first indication of wet AMD, straight lines may seem wavy. However, a dry AMD is caused by the macula gradually thinned with age, causing gradual blurring of central vision. More cases of AMD occur in the dry form, which progresses more slowly than the wet form. Dry AMD accounts for 70%–90% of AMD cases. Central vision gradually fades in the affected eyes as the macula is less active over time. Both the eyes are generally affected by dry AMD. Dry AMD is characterized by drusen which is one of its early symptoms.

1.4.2 Glaucoma

Glaucoma is a set of disorders that affects the optic nerve of an eye, resulting in blindness. The normal fluid pressure inside the eyes begins to rise as glaucoma progresses. Hence it is a word that refers to an eye disorder characterized by progressive optic nerve damage with a huge rise in IOP. The pressure inside the eye rises when there is an imbalance caused by either overproduction or incorrect drainage of the fluid. Thus, increased pressure damages the optic nerve when fluid accumulates as a result of poor drainage or surplus of fluid. The optic nerve is damaged by this pressure and hence glaucoma develops. The possible consequences of glaucoma are eye pain, hypertension, headache, redness in eye, inflammation of eye and blurred vision. Many major vision issues can be avoided with early detection and treatment.

The optic nerve has a normal disc and an indentation called the optic cup. Clinicians monitor the CDR and compare it with normal values to follow progression of glaucoma. Figure 1.5 shows a fundus image with glaucoma (Shahsuvaryan *et al.* 2005).

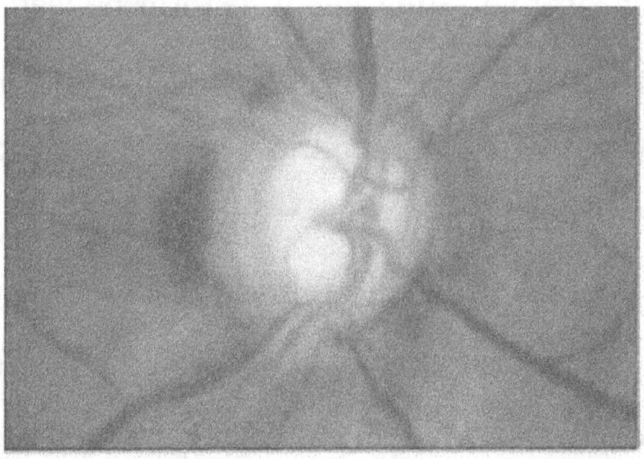

Figure 1.5 Cupping in Glaucoma

Glaucoma is divided into two types: (1) open angle glaucoma and (2) closed angle glaucoma. Open angle glaucoma, unlike the other eye disorders, grows slowly and it does not affect substantial vision loss till it is advanced; hence it is known as the "stealth thief of sight." Angle closure happens quickly and it might be painful. Despite the fact that visual damage may occur rapidly, individuals seek medical help before damage occurs as a result of the distress and pain.

1.4.3 Cataract

Cataract that clouds the lens of the eye is also one of the primary reasons of blindness. The aetiology of cataracts varies based on the patient's age. It may also be present at the time of birth. A majority of cataracts are characterized by age-related changes in eyes. Cataract harms one or both the eyes and develops slowly. Faded colors, blurring or double vision, difficulty in seeing in bright lights and difficulty seeing at night are some of its symptoms. Cataract results in having difficulty in driving, reading or recognizing faces. Cataract removal can be done at any time. Although there is no scientific evidence that wearing sunglasses (Neale *et al.* 2003) that block ultraviolet radiation can prevent cataracts, it may help to stop their progression. Figure 1.6 shows a fundus image with cataract where cloudiness is noticed (Hnoohom & Jitpattanakul 2017).

Figure 1.6 An Eye with Cataract

1.4.4 Diabetic Retinopathy

Diabetic retinopathy is an eye disease that affects people who have diabetes. Excessive amount of sugar harms the entire body, along with the eyes. Diabetes affects blood vessels all over the body, including retina, over time. The symptoms include noticing patches, hazy vision and a dark or empty place in the middle of the view. Figure 1.7 shows a fundus image with neovascularization (Shahsuvaryan *et al.* 2005). Neovascularization is the formation of new blood vessels in the optic disc or anywhere else on the retinal surface. Vision loss can be caused by damaged blood vessels or abnormal new ones. As diabetes can have an impact upon vision, it is extremely vital to obtain an eye checkup on a regular basis.

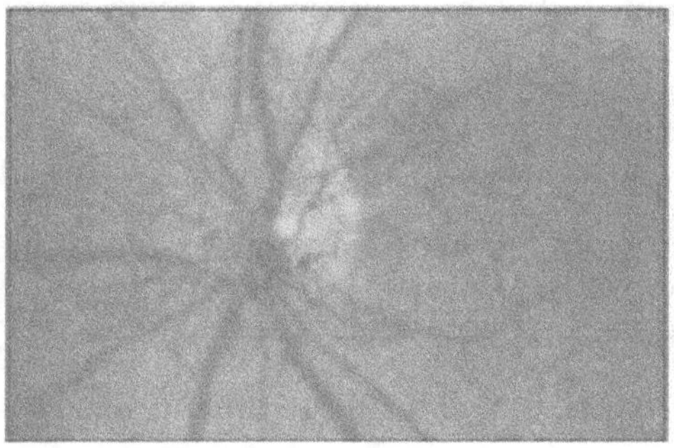

Figure 1.7 Neovascularization on the Optic Disc

Blood pressure, blood sugar and lipid problems must all be adequately controlled to lower the risk of diabetic retinopathy. Early detection and treatment, fortunately, lower the likelihood of irreversible visual loss.

1.4.5 Amblyopia

Amblyopia describes a condition in which one of the eyes is not functioning properly due to faulty brain function and poor vision. It is also known as "lazy eye". It is the most common form of vision impairment among children. It is a condition caused by different alignment of the eyes, strabismus, cataracts and so on.

1.4.6 Strabismus

Asymmetry between the two eyes is referred to as strabismus. It can produce esotropia (inward crossing of the eyes) or exotropia (outward turning of the eyes). In strabismus, the eyes do not focus on a single location at a time, but rather look in diverse directions. The cause of a child's strabismus is frequently unknown. There are many instances of congenital strabismus which is a condition that occurs shortly after birth or at the time of birth. Depth perception is reduced or eliminated when both the eyes are out of focus and the brain may learn to reject one eye's information, resulting in irreversible vision loss.

Glaucoma is the second leading cause for blindness in the world, after cataracts (Kingman 2004). The blindness it induces is permanent, so it may create a larger threat to the public than cataracts. Hence glaucoma is chosen for the current research.

1.5 PROBLEM STATEMENT

Image processing approaches are widely utilized in many fields. In recent years, the use of digital imaging in the medical field to diagnose a variety of disorders has grown in prominence. Following are the main issues being faced in fundus image processing:

- Glaucoma can be easily detected with the help of various image processing techniques. However, the classification accuracy still needs to be increased for the accurate glaucoma classification. Appropriate preprocessing, segmentation and feature extraction strategies are needed to develop an intelligent glaucoma classification system to diagnose glaucoma.
- Multiple investigations have been conducted on the image processing in the field of medical imaging. The health professionals and all the researchers continue to face some challenges even with the growing number of technologies

employed for fundus imaging and processing. Several methods have currently been used to solve these issues. A clear and precise result is required to choose the best strategy for processing a fundus image. Compared to all the other methods, machine learning methods have grasped the attention of the researchers in recent days.

- Machine learning methods, however, lack sufficient depth in expressing nonlinear functions as well as dealing with complex tasks. Deep learning based techniques offer an advantage over the conventional machine learning based techniques as they automatically extract the features. Moreover, they perform optimization in a way that they can respond really very well to high level of complexity.

1.6 RESEARCH MOTIVATION

Glaucoma is the second leading cause of blindness in the globe. Tham *et al.* (2014) have mentioned that 76 million people have been affected around the world. Glaucoma related blindness has been becoming increasingly common in India and around 11.9 million Indians have been affected by it. By 2040, it is estimated that 111.8 million people around the world will be infected with this disease. Hence its early detection is vital to avoid visual loss. Researchers are constantly striving to find more effective and safer treatments for this debilitating disease.

1.7 OBJECTIVES OF THE RESEARCH

The development of an automated intelligent computer based tool for the treatment of glaucoma is the main objective of the current research. There are four distinct objectives laid forth in the entire research.

- To investigate the publicly available data for glaucoma and to develop an intelligent classification system.
- To develop a novel segmentation and feature extraction algorithm for the detection of the glaucoma.
- To develop an automated system for classifying glaucoma disease based on machine learning and deep learning algorithms.
- To develop a novel ensemble CNN model that incorporates three deep learning models namely VGG16, Inception V3 and Inception-ResNet-V2 to improve the process of classification for detecting glaucoma.

1.8 PROPOSED METHODOLOGY

1.8.1 Glaucoma Classification System Using Machine Learning Algorithms

Machine learning and deep learning techniques are used widely in the majority of today's Artificial Intelligence tasks.

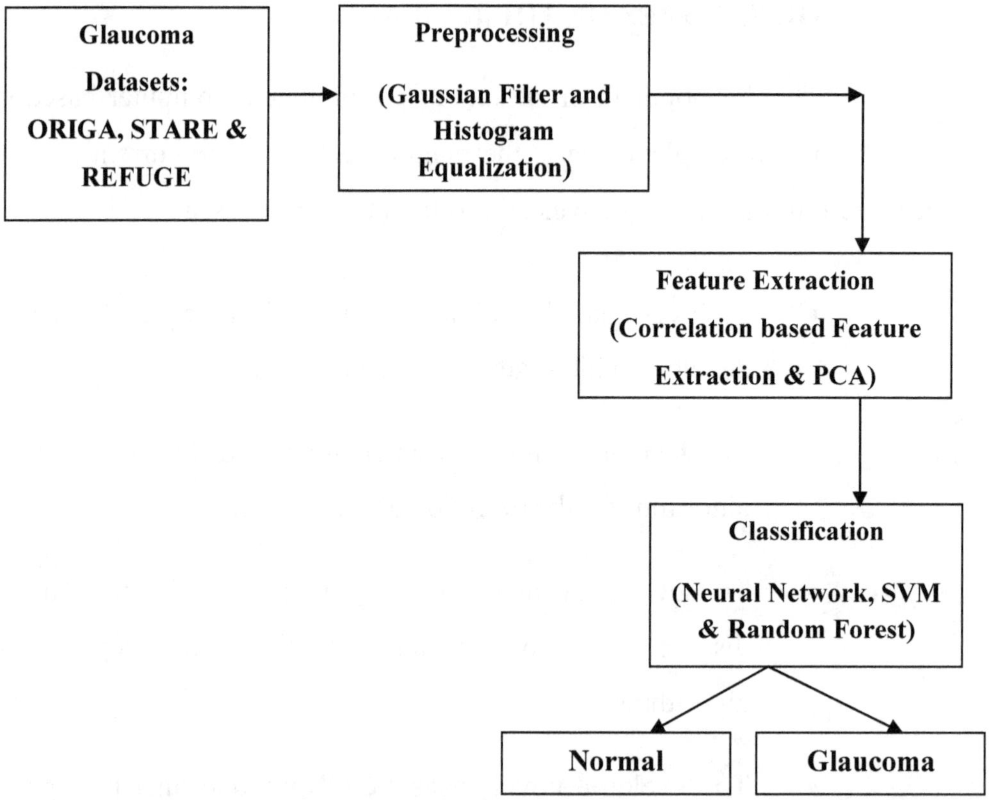

Figure 1.8 Classification of Glaucoma using Machine Learning

Figure 1.8 shows the classification of glaucoma using the conventional machine learning algorithms. In the current research, ORIGA, STARE and REFUGE which are the three publicly available datasets, are employed. The human retinal fundus images are initially preprocessed through Gaussian filter and Histogram equalization. Then image segmentation is done by using semantic segmentation approach. Features are extracted through density with correlation based method and the PCA is used to find the optimal features. The machine learning algorithms such as Neural Network, SVM and Random Forest are employed for the classification of glaucoma.

1.8.2 Glaucoma Classification System Using Deep Learning Algorithms

Figure 1.9 represents the block diagram for the classification of glaucoma using deep learning. The images from ORIGA, STARE and REFUGE datasets are given as the input to the deep learning models. The deep learning algorithms such as VGG16, Inception V3 and Inception-ResNet-V2 are employed to classify the images into either normal or glaucoma.

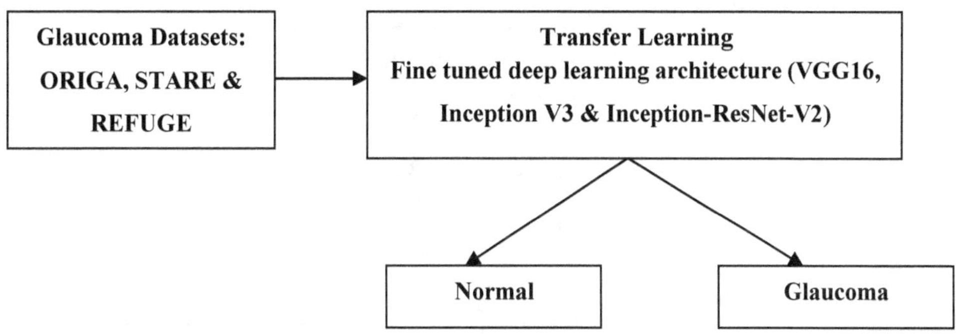

Figure 1.9 Classification of Glaucoma using Deep Learning

In the present research, an analysis has been proposed between machine learning and deep learning algorithms. The hyperparameters are tuned for both machine learning and deep learning algorithms. For learning, a pre-trained model has been utilized and the last layer of the model has been fine-tuned to fit it with the glaucoma classification. Learning rate and batch size have been updated to improve the performance of the model.

1.8.3 Glaucoma Classification System Using Ensemble Model

Glaucoma Classification using Ensemble Network (GCENet), a novel ensemble model that includes three deep learning models is developed to improve the classification accuracy for detecting glaucoma.

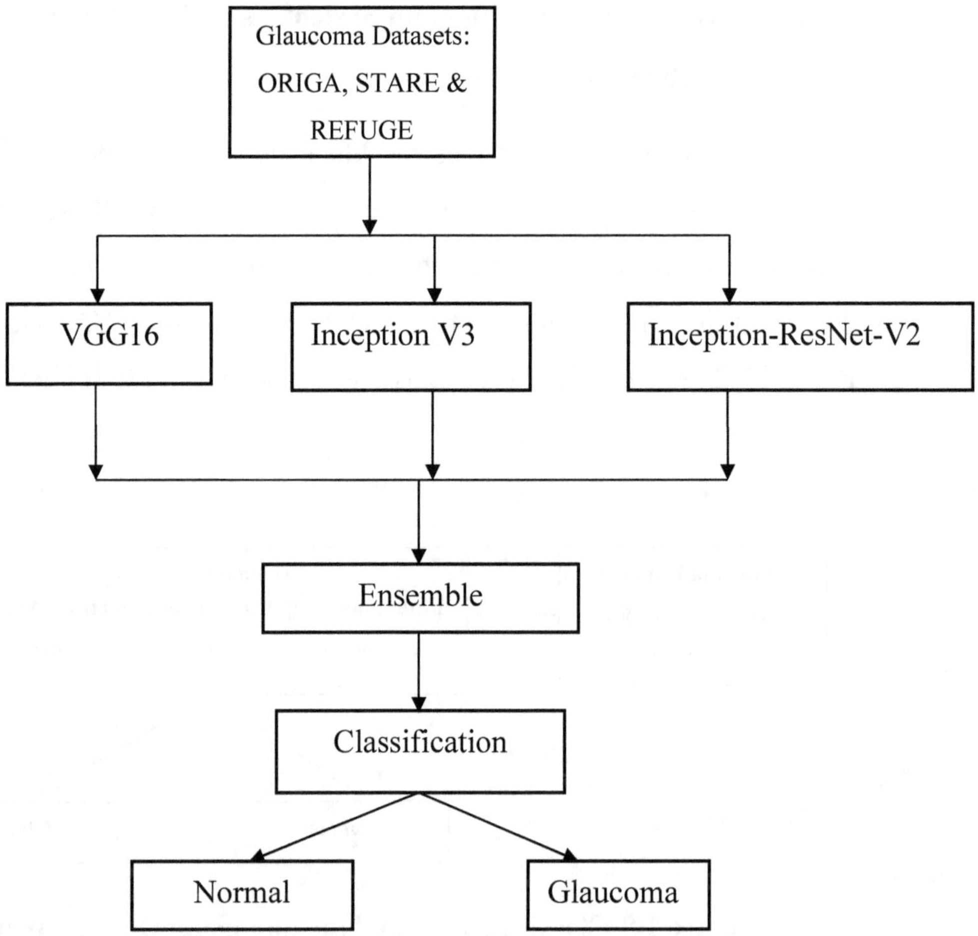

Figure 1.10 Classification of Glaucoma using Ensemble method

Figure 1.10 shows the classification of glaucoma using the ensemble model. In the current research, an ensemble of three distinct deep CNN models like VGG16, Inception V3 and Inception-ResNet-V2 is employed to detect glaucoma. The images from ORIGA, STARE and REFUGE datasets are given as the input to the deep learning architectures namely VGG16, Inception V3 and Inception-ResNet-V2. Later, the three distinct deep learning architectures are ensembled for the accurate classification of glaucoma.

The outcomes are recorded to meet the objectives as well as to research and explore novel data analysis strategies. The present research aims to construct an intelligent glaucoma classification model that uses machine learning and deep learning algorithms to detect chronic illness. This automated glaucoma classification system helps the doctors in assessing the risk of glaucoma in the patients.

1.9 ORGANIZATION OF THE THESIS

The thesis is divided into seven chapters.

Chapter 1 gives a brief overview of glaucoma, its diagnosis, various eye diseases, objectives and methodology of the current research.

Chapter 2 contains the literature review related to glaucoma. Various researches related to preprocessing methods, segmentation, feature extraction and classification algorithms employed for the detection of glaucoma have been reviewed in this chapter.

Chapter 3 deals with the dataset and the image preprocessing techniques. The general working of Gaussian filter and histogram equalization is also presented.

Chapter 4 outlines the segmentation and feature extraction phases. Various segmentation and feature extraction methods also have been presented.

Chapter 5 deals with the machine learning and deep learning algorithms employed for the classification of glaucoma. The performance of the proposed method has been analyzed.

Chapter 6 presents the outcome of the proposed model. The comparison of the proposed model with the other models also is made.

Chapter 7 includes the conclusion and future research directions related to the real-time glaucoma detection using computer assisted techniques.

CHAPTER 2

LITERATURE REVIEW

Medical imaging has grown rapidly in recent years for medical diagnosis. Image processing, Machine learning, deep learning and other approaches are used to detect glaucoma. The literature related to various preprocessing, segmentation, feature extraction and classification methods for the diagnosis of glaucoma is reviewed in this chapter.

2.1 GLAUCOMA

Glaucoma is a disease that damages the eye's optic nerves, causing vision loss or impaired vision. It is caused by low aqueous formation or blockage, which results in increased intraocular pressure and as a result, vision loss occurs. A prompt diagnosis is required to protect the eyes. Timely examination and treatment would avoid the loss of vision and the occurrence of various other diseases.

Medical imaging is an effective diagnostic tool for a variety of disorders. A number of imaging modalities have been invented. The use of imaging modalities expands the quantity of images that could be deciphered. The physicians are then faced with difficult conditions to obtain a more accurate diagnosis. CDR, disc diameter and RDAR are all considered clinically for diagnosing glaucoma.

Clinical authorities use CDR value to detect the progress of eye diseases. CDR should be less than 0.5 (Lim *et al.* 2007). Glaucoma can be inferred from a high cup to disc ratio. Manual subjective evaluation is tedious and time consuming, making it impractical for real-world applications. As a result, it is more feasible to create mathematical models that can consistently detect the disease. Computers are then employed to examine the medical images. The computerized objective evaluation that takes less time for computation would be a huge advance. As a result, an objective method is used to automate the prediction.

Glaucoma can now be identified due to the advancements in technologies. A lot of studies have recently been done using machine learning and deep learning approaches. For the early diagnosis of glaucoma, Weinreb *et al.* (2019) have used a retinal diagnostic imaging technology that allows clinical advisors to discriminate between glaucoma and non-glaucoma in a consistent manner for the benefit of the patients. AI based techniques for glaucoma detection are presented by Kaur & Khosla (2020). This study uses a variety of clinical, machine learning and deep learning methodologies employed for the detection of this devastating disease.

2.2 DATASETS

Various researchers have used different datasets for their research. A few datasets like ORIGA, STARE, RIM-1 r2, and DRIVE are used by recent researchers (Virbukaitė & Bernataviciene 2020). In the current research, ORIGA, STARE and REFUGE which are the publicly available datasets are used to train and test the model. Seventy percentage of data are utilized for training and thirty percentage of data are utilized for testing. There are 650 retinal images in ORIGA dataset which includes 482 healthy images and 168 glaucoma images. The STARE dataset has 50 images with glaucoma and 31 normal images. Likewise, the REFUGE dataset contains 1,200 retinal images which include 1,080 healthy images and 120 images with glaucoma. The commonly used datasets utilized by researchers to diagnose glaucoma has been shown in Table 2.1.

Table 2.1 Dataset for Glaucoma Diagnosis

Dataset	Without Glaucoma	With Glaucoma	Total Images
ORIGA (Zhang *et al.* 2010)	482	168	650
STARE (Hoover *et al.* 2003)	31	50	81
DRIVE (Staal *et al.* 2004)	33	7	40
ACRIMA (Diaz-Pinto *et al.*2019)	309	396	705
SINDI (Pan *et al.* 2011)	5,670	113	5,783
SCES (Sng *et al.* 2012)	1,630	46	1,676
ONHSD (Ko *et al.* 2016)	50	50	100
ARIA (Bankhead *et al.* 2012)	101	60	161
DRISHTI-GS1(Sivaswamy *et al.* 2014)	31	70	101
REFUGE (Orlando *et al.* 2020)	1,080	120	1,200
HRF (Odstrcilik *et al.* 2013)	18	27	45
RIM-ONE (Fumero *et al.* 2011)	92	39	131

2.3 IMAGE PROCESSING TECHNIQUES

There is a huge rise in the automated methods for testing a large number of people for the vision-threatening diseases such as glaucoma, diabetic retinopathy, cataract and so on. Recently, image processing has become a viable and helpful screening method. Ophthalmologists employ image processing to keep track of the development of reactions to therapy because it provides a high-quality everlasting record of the fundus images. A large number of studies have been done on glaucoma detection.

2.3.1 Image Preprocessing

Preprocessing is an essential step, as features retrieved from a preprocessed image are more optimal than those taken from a raw image. It is a vital step in achieving high accuracy. It is used to improve the quality of the image by removing undesired distortions and increasing visual attributes that are crucial for subsequent processing of the image. Improving the quality of the images would have a major impact upon the performance and efficiency of the proposed method. The preprocessing approach can be used to attain this goal. Hence, Gaussian filter and Histogram Equalization are used in this current research. There are a few distinct methods like Filtering with morphological operators, Linear contrast adjustment, Noise removal using Wiener filter, Median filtering, Unsharp mask filtering, Decorrelation stretch and Contrast-limited Adaptive Histogram Equalization (CLAHE) used for the preprocessing of the images. These methods help to eliminate noise, sharpen or brighten an image to make it simpler to find significant features.

There are various preprocessing methods used by researchers for improving the quality. Acharya *et al.* (2017) have implemented a computerized diagnostic solution using private datasets. Adaptive histogram equalization is employed to enhance the contrast in images. Textons indicate the basic elements in the image. To extract the textons, the gray-scale images are convolved with the 4 different filter banks such as Leung-Malik (LM), Maximum Response 4 (MR4), Maximum Response 8 (MR8) and Schmid. These textons are used to extract Local Configuration Pattern (LCP) features.

To identify unique features and to discard the repetitive features, the Sequential Floating Forward Search (SFFS) method is paired with t-test. The kNN classifier employed for the LM filter bank has obtained the maximum accuracy of 95.8%, sensitivity of 96.7% and specificity of 92.3%.

A new automatic diagnostic system has been designed by Soltani *et al.* (2018). Gray-scale conversion, histogram equalization and noise filtration are employed as preprocessing steps. The Gray scale conversion is employed to convert the pixels from a three-dimensional plane image to a one-dimensional plane image. Further, the adaptive histogram equalization technique is used to distribute gray tones consistently over the image. The non-uniform illumination and pigment color difference in the retinal fundus image can be removed using this procedure. The adaptive histogram equalization technique entails computing histograms of various portions of the image and applying them to the output image to achieve a uniform distribution of pixel intensities. It features a decision making framework based on fuzzy logic approaches that provide accurate and trustworthy diagnosis to medical personnel. Bilateral as well as Gaussian filters are used at first to attenuate high frequencies that relate to noise. These filters preserve the discontinuities. Results prove to be superior to the other existing models. The percentage of correct predictions is greater than 96 and an accuracy improvement of 1–9% than the previous techniques has been achieved.

Similarly, Li *et al.* (2018) have investigated the effectiveness of deep learning methods for detecting Glaucomatous Optic Neuropathy. A total of 48,116 fundus images are investigated. The ophthalmologists have labeled the images. Before the feature extraction, the image have been enhanced and normalized. During the training, the methods like dropout as well as regularization are also applied. The Area Under the Curve (AUC) of 0.986, sensitivity of 95.6% and specificity of 92% have been achieved.

Panda *et al.* (2018) have employed the CLAHE technique for preprocessing images to overcome the difficulties of fundus images. An automatic approach is created for detecting Retinal Nerve Fiber Layer Defect (RNFLD). It is examined by utilizing the imaging modalities like OCT and Scanning Laser Polarimetry (GDx), which are extremely expensive in everyday practice. Local minima analysis of 1-D intensity profiles on concentric circles identifies the initial border pixels. The real border pixels are recognized using a random forest with the suggested Cumulative Zero Count Local Binary Pattern (CZC-LBP) and Directional Differential Energy (DDE) containing Shannon, Tsallis entropy and intensity characteristics. The AUC of 0.8733 and the sensitivity of 78.21% have been obtained.

Sahu *et al.* (2019) have employed the CLAHE approach to tackle the difficulties involved in denoising and color enrichment in fundus images. The STARE database is used. For de-noising the fundus image, different filters such as median, Gaussian, Wiener, average and weighted median are applied individually. With the CLAHE approach, the low contrast is improved. Compared to the state-of-the-art method, this proposed method produces 7.85% of increase in Peak Signal to Noise Ratio (PSNR), 1.19% of increase in Structural Similarity Index Measure (SSIM), 0.12% of improvement in Correlation Coefficient (CoC) and 1.28% of improvement in Edge Preservation Index (EPI).

In the same way, Abdel-Hamid (2020) has presented a wavelet based glaucoma predictive model. Various improvement approaches are used to improve the red channel's capability for detecting optic disc. The preprocessing methods like unsharp masking as well as CLAHE are employed to improve the red channel's overall contrast and to increase the appearance of the optic disc against the image background. AUC of 96.7% and the accuracy of 96.7% have been achieved.

Tekouabou *et al.* (2022) have presented a combined model with preprocessing approaches and ensemble classifier for the diagnosis of glaucoma. Linear SVM has been employed for selecting the relevant features. Two sets of private datasets are utilized. The Bagging Classifier (BC) has been associated with Dynamic/Static Selection Strategies (BC-DS/SS) in order to enhance the performance of the model and the maximum accuracy of 97% has been achieved.

2.3.2 Image Segmentation

Image segmentation is an important concept in image processing that has a wide range of applications. Many image segmentation algorithms have been developed by various researchers.

Welfer *et al.* (2010) have proposed a new adaptive technique for optic disc automated segmentation by employing mathematical morphology to detect significant features of optic disc like the optic disc locus and the optic disc rim. The two publicly available datasets DRIVE and DIARETDB1 are used. The model is evaluated on 40 images from the DRIVE database and 89 images from the DIARETDB1 dataset. The DRIVE and DIARETDB1 databases have 100% and 97.75% of success rates respectively in optic disc localization.

The detection of the optic disc is a crucial step in the development of systems for the automated diagnosis of a variety of significant ocular illnesses. Aquino *et al.* (2010) have suggested a new optic disc analysis model. The optic disc is segmented from computerized retinal images using a new template based technology. To create a circular optic disc border approximation, this technology employs morphological as well as edge detection techniques, accompanied by Circular Hough Transform. The

MESSIDOR database, which is available freely, is used to test the methods. In 99% of the cases, localization has been effective and in 86% of the cases, segmentation has been efficient.

Segmentation is more complex because of the great volume of blood vessels in the optic cup. Furthermore, the progressive shift in the strength between the neuroretinal rim and the cup complicates cup segmentation. Glaucoma also affects the curvature of the cup area. Damon et al. (2012) have presented a method for automatically detecting the optic cup edge using kinks. Within the optic disc, patches are identified from which segment based wavelet, edge and color features for vessel candidates are obtained. SVM algorithm is utilized for categorization. To determine the optic cup, the kinks derived are coupled with pallor based information. The cup width discrepancy has been reduced by 43.3% when kinks are used in the identification of the cup.

Cheng *et al.* (2013) have proposed super pixel classification method for detecting glaucoma in the optic cup. Each optic disc image is initially over-segmented into super pixels in the suggested method. The initial step is to segment the optic disc using the super pixel classification model. For classification, the features like content, shape and colors are taken from the super pixel. To categorize each super pixel into cup or non-cup, mean intensities, centre surround statistics and location information are extracted. The 650 images from ORIGA dataset and 1,676 images from Singapore Chinese Eye Study (SCES) dataset are used to test the suggested method. Experimental results show an average overlapping error of roughly 26.0% which is substantially better than the previous approaches. This approach could be used to diagnose glaucoma.

Similarly, Zhao *et al.* (2015) have applied active contour model for the segmentation of vessels. This model employs the combination of intensity information and local phase based improvement map for enhanced segmentation. It is tested on the publicly available DRIVE and STARE datasets. For the STARE dataset, a maximum accuracy of 95.6%, sensitivity of 78% and specificity of 97.8% are obtained. However, for the DRIVE dataset, the accuracy of 95.4%, sensitivity of 74.2% and specificity of 98.2% are achieved.

Imani *et al.* (2015) have implemented Morphological Component Analysis (MCA) method with appropriate transforms to differentiate vessels from lesions. The model is tested with DRIVE and STARE datasets. For the DRIVE dataset, the accuracy of 95.23%, sensitivity of 75.24% and specificity of 97.53% are achieved. In contrast, for the STARE dataset, the accuracy of 95.90%, sensitivity of 75.02% and specificity of 97.45% are accomplished.

Similarly, Wang *et al.* (2015) have employed CNN and Random Forest to solve the challenge of retinal blood vessel segmentation. The DRIVE and STARE datasets are used to test the model. The system learns features from images and predicts the patterns. For the DRIVE dataset, the accuracy of 97.67%, sensitivity of 81.73% and specificity of 97.33% are obtained. Likewise, for the STARE dataset, the accuracy of 98.13%, sensitivity of 81.04% and specificity of 97.91% are obtained.

Kaur *et al.* (2017) have applied Matched Filter and its First Derivative of Gaussian (MF-FDOG) for image segmentation and unsharp masking as a preprocessing technique to increase the quality of the image. Both the private and the publicly available datasets like DRIVE, STARE, ARIA and HRF are utilized. Neural network based classifier is used for classification. It produces the accuracy of 95.91%, sensitivity of 83.06% and specificity of 97.30%.

Fu *et al.* (2018) have proposed M-Net to segment both the optic disc and the optic cup. The ORIGA and SCES datasets are used. A multilabel loss function is suggested to achieve the final segmentation result. To improve the performance of the joint optic disc and optic cup segmentation, Polar Transformation has been used. On both the datasets, the CDR is determined, resulting in effective development for glaucoma diagnosis.

In the same way, Al-Bander *et al.* (2018) have recommended Fully Convolutional DenseNet for segmentation. CDR has been assessed by segmenting the optic disc and the cup. The result of the Optic Disc Diameter (ODD) calculation, which is used to diagnose glaucoma, is used to determine the Region of Interest (ROI). In the study, the model is evaluated using the five different datasets such as ORIGA, ONHSD, RIM-ONE, DRIONS-DB and Drishti-GS. The network is built on a single dataset before refining it on four further datasets. For the new test data, the model performs well. The accuracy of 99.89% for ORIGA, 99.66% for DRIONS-DB, 99.69% for Drishti-GS, 99.9% for ONHSD and 99.22% for RIM-ONE has been obtained.

Yun *et al.* (2019) have recommended GL-Net in conjunction with the GAN technique for the segmentation of both the optic disc and the optic cup from DRISHTI-GS1 dataset. To address the issue of insufficient data and overfitting during the training process, transfer learning and augmentation are used. In this work, VGG 16 is employed. Adam Optimizer is used for optimization during the training. It has been indicated that this proposed GL-Net is effective in diagnosing glaucoma by segmenting both the optic disc and the optic cup.

Guo *et al.* (2020) also have proposed an automated method of diagnosing glaucoma. The UNet++ network has been enhanced for segmentation. After segmentation, the CDR is measured as a crucial parameter. Increasing Field of View (IFOV) is utilized to perform the necessary extraction. It also incorporates Adaptive Synthetic (ADASYN) which is used to resolve data imbalance issues. The research is conducted using the DRISHTI-GS1 and ORIGA datasets. This work achieves the accuracy of 84.3%, sensitivity of 89.4 and 0.901 of AUC.

Tabassum *et al.* (2020) have used Cup Disc Encoder Decoder Network (CDED-Net) on the three datasets such as REFUGE, RIM-ONE and DRISHTI-GS. Drishti for optic cup has produced a sensitivity of 95.67% and specificity of 99.81%. Drishti for optic disc has given a sensitivity of 97.54% and specificity of 99.73%. Moreover, Rim-one for optic cup has produced a sensitivity of 95.17% and specificity of 99.81%. However, the Rim-one for optic disc has given a sensitivity of 97.34% and specificity of 99.73%. REFUGE has obtained 0.8837 of Intersection over Union (IoU) for optic disc, 0.8111 of IoU for optic cup and 0.8705 of Mean Intersection over Union (MIoU).

Ali *et al.* (2020) have employed one public dataset RIM-ONE-r3 and one private dataset for screening glaucoma. Fuzzy Broad Learning System (FBLS) has been suggested for segmenting the optic disc and the optic cup. As a result, an AUC of 0.860 has been accomplished. It has been concluded that this approach outperforms the other existing systems.

Afolabi *et al.* (2021) have presented a U-Net lite, an updated U-Net model and Extreme Gradient Boost (XGB) algorithm for the detection of glaucoma. The optic disc and the cup are extracted from the fundus images using the suggested model. After analyzing the features extracted from the segmented optic disc and cup, the XGB algorithm is used to identify glaucoma. The suggested U-Net lite model is trained and tested on the four datasets, namely DRISHTI-GS, DRIONS, RIM-ONE V2 and RIM-ONE V3. The model obtains the average dice-scores of 0.97 on DRISHTI-GS and RIM-ONE V2 and 0.96 on RIM-ONE V3 and DRIONS for optic disc segmentation. The XGB algorithm diagnoses glaucoma with the accuracy of 88.6% and AUC-ROC of 93.6% from DRISHTI-GS and RIM-ONE V3 datasets.

Lei *et al.* (2021) have presented an Image Synthesis and Feature Alignment (ISFA) approach for segmenting optic cup and optic disc from the retinal fundus images. REFUGE dataset is utilized for training the model. Further, RIM-ONE_r3, ORIGA and Drishti-GS datasets are employed for testing. To provide feature consistency for the images, Content and Style Feature Alignment (CSFA) algorithm has been utilized. An adaptive learning has been employed for retrieving the domain constant features. An Edge Attention Module (EAM) also has been developed for low-level feature maps in order to improve the representational capability of domain constant edge structure information. The method has given the maximum Dice similarity coefficient of 0.96 in the Drishti-GS dataset for the optic disc segmentation. It has been concluded that the segmentation outcomes achieved by the proposed algorithm are better suited to help clinicians to diagnose glaucoma.

2.3.3 Feature Extraction

Feature extraction is a process of representing large amounts of data in a compact way such that the subsequent computations are simple and effective. Various feature extraction methods have been employed by many researchers.

Dua *et al.* (2011) have extensively explored texture features inside images for precise and efficient classification of glaucoma. To identify the crucial texture properties, wavelet subbands have been used. The selective capability of wavelet features produced from daubechies (db3), biorthogonal (bio3.3, bio3.5, and bio3.7) and symlets (sym3) wavelet filters are investigated in the study. A new method has been presented for extracting energy signatures from 2-D discrete wavelet transforms and then it has been subjected to various feature rank as well as feature selection algorithms. Using SVM, Random Forest, sequential minimum optimization and Naive Bayes classification techniques, the performance has been evaluated. To prove the efficiency of these methods, tenfold cross-validation is employed and an accuracy of about 93% has been achieved.

Optical coherence tomography, Heidelberg retina tomography and scanning laser polarimetry are employed to evaluate the retinal optic nerve fiber layer. Acharya *et al.* (2011) have described a novel glaucoma detection approach based on a mix of texture as well as Higher Order Spectra (HOS) data from digital fundus images. Supervised classification, sequential minimum optimization, naive Bayesian, SVM and Random Forest classifiers also are utilized. It has been found that the texture and HOS features, z-score normalization and feature selection, perform much better than the other classifiers. To improve the result, the significance of feature ranking as well as normalization is also explored. It accurately identifies glaucoma images with an efficiency of more than 91% when paired with a random forest classifier.

Kim *et al.* (2013) have explored the characteristics of Fractal Analysis (FA) to categorize glaucoma detection into multiple classes. The features are generated and exposed to feature selection by using PCA and Gaussian kernel approach. The evaluation metrics for the FA features and

Wavelet-Fourier Analysis (WFA) including Fast-Fourier Analysis (FFA) are computed. In the study, 90% of the images are employed for training and 10% of the images are used for testing. The classification accuracy of 82%, 86% and 88% has been achieved for WFA, FFA and FA respectively.

Singh *et al.* (2015) have suggested a glaucoma diagnosis approach based on automatic image processing with a digital fundus image. Wavelet feature extraction has been employed. It is then followed by efficient genetic feature selection. It is clear from the experiment that for the identification of glaucoma with fundus images, the wavelet characteristics of the fragmented optic disc are clinically more important than the characteristics of whole or subdivided fundus image. The efficiency of glaucoma identification attained is 94.7% and a comparison of this method to the existing methods of glaucoma disease diagnosis using fundus images has shown that this suggested method has given improved classification accuracy.

Furthermore, Shehryar *et al.* (2017) have developed an automated approach for detecting glaucoma. The private dataset is used for the study. Histogram equalization is utilized to preprocess the data to increase the contrast of the images. Hybrid Structural Features (HSF) and Hybrid Textural Features (HTF) also have been utilized to extract the features from the optic disc and the cup segments. Further, SVM is used to classify the glaucoma into three categories: normal, glaucoma and suspicious. HSF produces the accuracy of 83%, sensitivity of 77% and specificity of 88%. In contrast, HTF obtains the accuracy of 94%, sensitivity of 96% and specificity of 92%.

Raghavendra *et al.* (2017) have used a non-parametric Global Image descriptor (GIST) and optic disc localization to construct a glaucoma diagnosis model. The method suggests segmentation of the optic disc before applying the Radon Transformation (RT). Modified Census Transformation

(MCT) compensates for the variation in the illuminance of the radon converted image. Additionally, GIST descriptors are used to extract features. Locality Sensitive Discriminant Analysis (LSDA) also is employed to lower the GIST descriptor's dimension, which is then followed by various feature selection and ranking algorithms. The rated features are used to create an effective glaucoma classifier. Using SVM, the accuracy of 97%, sensitivity of 97.80% and specificity of 95.80% have been achieved.

Chai *et al.* (2018) have designed the MultiBranch Neural Network (MB-NN) to select important areas from retinal images. Through the development of deep learning algorithms, the study focuses on the diagnosis of glaucoma based on a combination of domain knowledge and hidden features. The MB-NN has been proved to be of higher quality with the accuracy of 91.51%, sensitivity of 92.33% and specificity of 90.90%.

In the same way, Convolutional Neural Network (CNN) has been proposed by Raja *et al.* (2020) for automated segmentation to identify glaucoma. Layer pixels are retrieved using structure tensors in the suggested system and VGG16 is used for the two most critical aspects like i) Feature Extraction and ii) Classification. CDR is measured to diagnose glaucoma. The accuracy of 94.68%, sensitivity of 94.6% and specificity of 94.07% have been accomplished and so the suggested method has been considered to be a reliable one.

Song *et al.* (2021) have suggested an efficient, reliable and automated approach for glaucoma diagnosis. The private dataset gathered from the Kaohsiung Chang Gung Memorial (KCGM) Hospital, Taiwan, is used. The dataset is divided into 50%, 20 % and 30% and employed for training, validation and testing process respectively. CNN is used in the proposed method along with the proposed generalized loss function, Retinex theory and reliable Design of Experiments (DOE). The proposed model has been proved to be more efficient for glaucoma diagnosis. It obtains 95% of sensitivity, 98% of specificity and 97% of accuracy.

Nayak *et al.* (2021) have proposed a model named Evolutionary Convolutional Network (ECNet) to extract the features from the retinal fundus images and to diagnose glaucoma. The private dataset gathered from the Kasturba Medical College, Manipal, is utilized. The layers like convolutional, Rectified Linear Unit (ReLU), compression and summation are employed to extract discriminative features easily. To tune the weights at various layers, a Real-Coded Genetic Algorithm (RCGA) is used. Further, to choose the best performing model, the resulting feature vectors are then processed through a number of classifiers like K-Nearest Neighbour (KNN), SVM, Backpropagation Neural Network (BPNN), Extreme Learning Machine (ELM) and Kernel ELM (K-ELM). The proposed method obtains the maximum accuracy of 97.20 % for the SVM classifier.

Chen *et al.* (2022) have recommended the deep learning method for the diagnosis of glaucoma. Variational AutoEncoder (VAE) architecture is employed for training and the deep learning method is used to determine the correlation in the images and the RNFL thickness distribution. Private dataset is utilized. The performance of the U-Net model at various depths has been assessed. PSNR is observed to have been improved with the depth of the network design. During training and validation, the PSNR values range as 23.8 dB and 18.8 dB in a 4X4 configuration and 27.6 dB and 19.4 dB in a 6X6 configuration, respectively.

Furthermore, Mahmood & Lee (2022) have proposed an approach for precise optic disc localization and detection from the fundus images by analyzing the color and blur features. In the proposed approach, the input fundus image is transformed into color space in order to improve the contrast and the shape of the optic disc. Further, the extensive maxima transformation and the accumulated directional blur are used to extract the optic disc effectively. Later, the optic disc is distinguished by applying the radial blur to generate superior profiles. Ultimately, the optic disc is selected using the full width at 80% of highest (FW80M) measurement.

2.3.4 Image Classification

Classification is a method of classifying images by extracting features and assigning a label to them. Compared to the conventional machine learning and statistical methods, the deep learning methods like CNNs have made significant advancements in computer vision in recent years.

CNN is being used for a lot of image classification tasks. Image classification has become a lot easier with the help of CNN. CNN consists of many convolutional layers, pooling layers and fully connected layers. To extract features, convolutions and pooling layers are used. A convolution filter is used to apply convolution to the input image to generate a feature map and the output dimensionality is reduced by the pooling layers. Fully connected layers are used to develop a classifier. The framework is provided with the images and the result is analyzed after forward propagation. Back propagation updates weights, which improves learning. This, in turn, helps to improve the accuracy of the model.

Asaoka *et al.* (2017) have employed random forest classification on the private dataset for the identification of glaucoma. The Area under the Receiver Operating Characteristic curve (AROC) metric is used to assess the outcome. Result shows that 93% of AROC has been obtained using Random Forest approach. This result has been compared with the other models like the stepwise model selection, Least Absolute Shrinkage and Selection Operator (LASSO) model and ridge model. It has been concluded that this proposed model significantly performs well than those existing models. AROC of 71.9% has been achieved in the stepwise model selection. LASSO gives 89.6% of AROC while the ridge model produces 89.2% of AROC.

Fu *et al.* (2018) have used the MultiContext Deep Network (MCDN) on the AS-OCT (Anterior Segment OCT) dataset to assess Angle Closure Glaucoma. Data augmentation also is used to increase the training data count. The integration of the clinical features identified results in improved performance. The implementation is carried in Tensorflow and VGG-16 classifier model has been used. The accuracy of 89.26%, AUC of 0.9456, sensitivity of 88.89% and specificity of 89.63% have been obtained.

Raghavendra *et al.* (2018) also have proposed a unique Computer-Aided Diagnosis (CAD) with eighteen layers of CNN for diagnosing glaucoma. The images are taken from Kasturba Medical College, Manipal. For training and testing, 70% and 30% of the data are used respectively. The accuracy of 98.13%, sensitivity of 98% and specificity of 98.30% have been obtained.

Christopher *et al.* (2018) have applied the deep learning method for the diagnosis of glaucoma using private datasets. Experimentation is done using the transfer learning models such as VGG16, ResNet50 and Inception v3. ResNet50 has given the accuracy with AUC of 0.91, sensitivity of 88% and specificity of 95%.

Similarly, Ahn *et al.* (2018) have used deep learning framework for the detection of glaucoma. The private dataset contains 1,542 fundus images, out of which 754 images are used in the training phase, 324 images are employed in the validation process and 464 images are used for the testing phase. In both logistic classification and CNN, accuracy has been measured. The logistic classification has produced 82.9%, 79.9%, and 77.2% of training, validation and testing accuracies respectively. Moreover, CNN has given the training, validation and testing accuracies of 92.2%, 88.6% and 87.9% respectively.

Singh *et al.* (2019) have used the transfer learning method to predict glaucoma. On RIM-ONE, Inception-v3 has achieved the accuracy of 92.51%. Data augmentation and preprocessing techniques like normalization and CLAHE are used. Adam optimizer also is used to optimize the weights. Binary cross entrophy is employed as a loss function. A total of 20 epocs are used to obtain the training accuracy of 91.54% and the validation accuracy of 90.91%. The testing accuracy of 92.51% also has been achieved for the test data.

Gomez-Valverde *et al.* (2019) have proposed the CNN approach with transfer learning for diagnosing glaucoma. The experiment is carried out on three different datasets, where 2,313 images are used for the training, validation, and testing processes. Using the VGG19 model as a transfer learning, the best result of 0.94 of AUC has been obtained. Finally, color fundus images are utilized to evaluate the data integration capability with medical records and tonometry tests.

Raja *et al.* (2020) have used Hybrid Convolutional Network on the dataset named Armed Forces Institute of Ophthalmology (AFIO). The F1 Score of 0.9577 and the accuracy of 91.17% have been obtained. The most important aspect is that the system developed diagnoses the severity and gives the solution that is appropriately validated on a standard dataset. It has been clinically proven for the identification of glaucoma.

Li *et al.* (2020) also have applied a new deep learning method and Attention based CNN (AG-CNN) for diagnosing Glaucoma. Large-scale Attention based Glaucoma (LAG) dataset and public dataset (RIM-ONE) are used. For RIM-ONE dataset, the accuracy of 85.2%, sensitivity of 84.8% and specificity of 85.5% have been obtained and for LAG dataset, the accuracy of 96.2%, sensitivity of 95.4% and specificity of 96.7% have been achieved. It is proven that this method is more trustworthy and it produces superior results. It is noted that the image redundancy also has been reduced and this helps to improve the accuracy.

Zulfira *et al.* (2021) have proposed the Dynamic Ensemble Selection (DES) to classify glaucoma with severity. Active Contour Snake (ACS) is employed to compute CDR. Further, Peripapillary Atrophy (PPA) is recognized using segmentation and thresholding algorithms and the features are then retrieved using a Gray-Level Co-occurrence Matrix (GLCM). The model has been tested with the three datasets, namely RIM-ONE, MESSIDOR and images from KAGGLE. The proposed model achieves a greater mean accuracy of 0.96.

Furthermore, Devecioglu et al., (2021) have recommended compact Self-Organized Operational Neural Networks (Self-ONNs) for glaucoma diagnosis. Three datasets, namely ACRIMA, RIM-ONE and ESOGU are employed in the research. For all the three datasets, the suggested classifier outperforms with 8–12% of F1 score compared to the other deep CNN frameworks.

Ajesh *et al.* (2021) have utilized Deep Belief Network (DBN) with multi-feature analysis for the glaucoma diagnosis. MESSIDOR dataset has been used and it is separated into three sections: training (70%), testing (20%) and validation (10%). The accuracy of 95.9%, sensitivity of 88.3% and specificity of 95.12% have been achieved. In addition to the proposed approach, these metrics are compared with several other algorithms such as SVM, Morphological Reconstruction and Empirical Wavelet Transform using SVM. It has been proved that the proposed approach has produced superior outcomes.

Islam *et al.* (2021) also have employed different deep learning approaches like EfficientNet, GoogLeNet, DenseNet and MobileNet for the detection of glaucoma. Both the private and the public datasets have been used. The private datasets are gathered from Bangladesh Eye Hospital, Dhaka (BEH) and ACRIMA is the public dataset employed. From these datasets, blood vessels, optic disc and optic cup are segmented and for the cropped optic cup and disc fundus images, the EffcientNet-b3 architecture produces the best results, with test accuracy of 0.9652, F1-score of 0.9512 and ROC AUC value of 0.9574.

2.4 RESEARCH GAP

From the literature review, it is inferred that an extensive research has been carried out in the diagnosis of glaucoma. Various researchers have explored numerous methods for reducing the risk of glaucoma in patients at an early stage. The methods developed have their own limitations which are highlighted below.

- Appropriate preprocessing strategies are needed for accurate classification of glaucoma.
- Segmentation and feature extraction are difficult tasks that demand a comprehensive and accurate approach. Due to the obvious existence of blood vessels, accurate segmentation becomes a time-

consuming process. To obtain accurate segmentation and to extract features from the images, an effective method is required.

- The classification accuracy still needs to be increased for diagnosing glaucoma. A clear and precise result is required to choose the best strategy for processing a fundus image. Compared to all the other methods, machine learning and deep learning methods have grasped the attention of the researchers in the present situation.

2.5 SUMMARY

The existing techniques used for the detection of normal and anomalous occurrences are mostly addressed in the literature. Machine learning algorithms are commonly used in the existing glaucoma diagnosis systems. Effective segmentation and feature extraction algorithms are required to diagnose early ophthalmic illnesses. Results appear to imply that the preprocessing methods, segmentation and features chosen, classification algorithm, all have an impact upon the final results, therefore all the four phases are equally important. Various studies with respect to above aspects for the detection of glaucoma have been reviewed in this chapter.

CHAPTER 3

GLAUCOMA IMAGE ACQUISITION AND PREPROCESSING METHODS

Data collection is the first stage in any diagnosis using medical applications. An original raw image collected from the patients requires preprocessing methods to make it reliable for the specific applications. Data preprocessing can organize this unstructured data to structured data. Once the data are organized, they are converted into the specified format required for the machine learning algorithms to construct the model.

3.1 DATA COLLECTION

The retinal fundus image is a great resource for evaluating retinal health and commonly used to diagnose ocular illnesses. The dataset allows the researchers to assess their computer-aided techniques. Collection of glaucoma images from the common datasets is the first step in detecting glaucoma. The datasets consist of both healthy eye and glaucoma affected eye and have been labeled by field experts. Figure 3.1 shows the normal eye and the glaucoma affected eye.

A vast amount of datasets can be utilized to investigate the presence of glaucoma using the machine learning and deep learning methods. Various researchers have used different datasets for their research. A few datasets like ORIGA, STARE, RIM-1 r2 and DRIVE are used by recent research (Virbukaitė & Bernataviciene 2020).

REFUGE dataset contains many images and used by many researchers. Hence for the current research, ORIGA, STARE and REFUGE datasets are selected and these are the publicly available datasets.

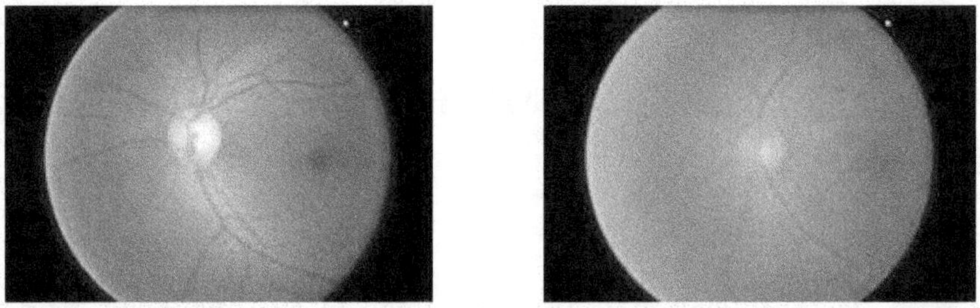

a) Normal Eye b) Glaucoma affected eye

Figure 3.1 Normal eye and Glaucoma affected eye

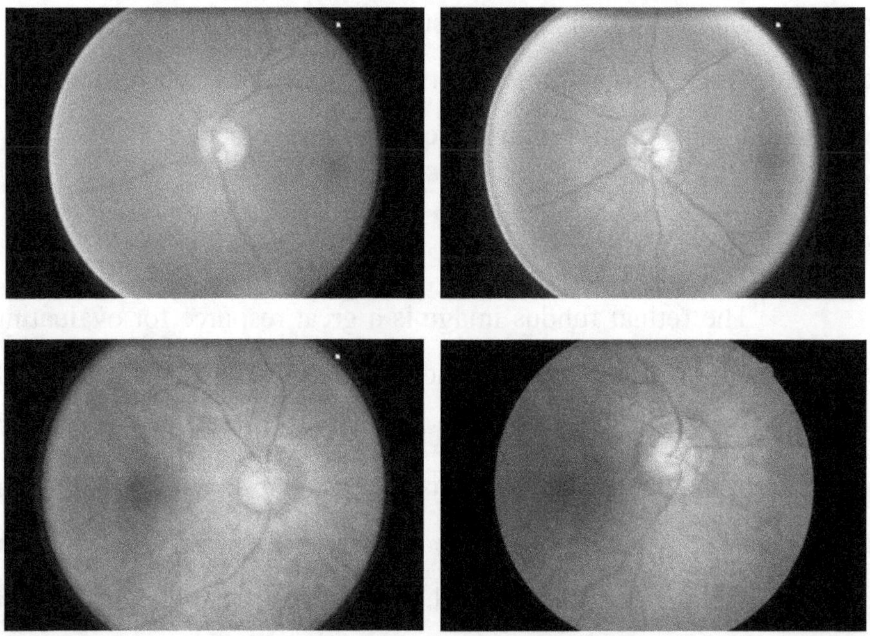

Figure 3.2 Retinal fundus images from ORIGA Dataset

The ORIGA dataset contains 650 retinal fundus images with the resolution of 3072 × 2048 pixels, with the labels assigned by experts. It has 482 healthy images and 168 images with glaucoma. It has been gathered from the Singapore Malay Eye Study (SiMES) managed by the Singapore Eye

Research Institute (SERI). Figure 3.2 shows the retinal fundus images from ORIGA dataset.

The STARE is an openly accessible dataset. It has 50 images with glaucoma and 31 normal images. In it, the images have been taken in a camera with a 35-degree field of view and with a resolution of 700 x 605 pixels. It is being used by many researchers who focus on analysis of images for diagnosing diseases. Figure 3.3 shows the retinal fundus images from STARE dataset.

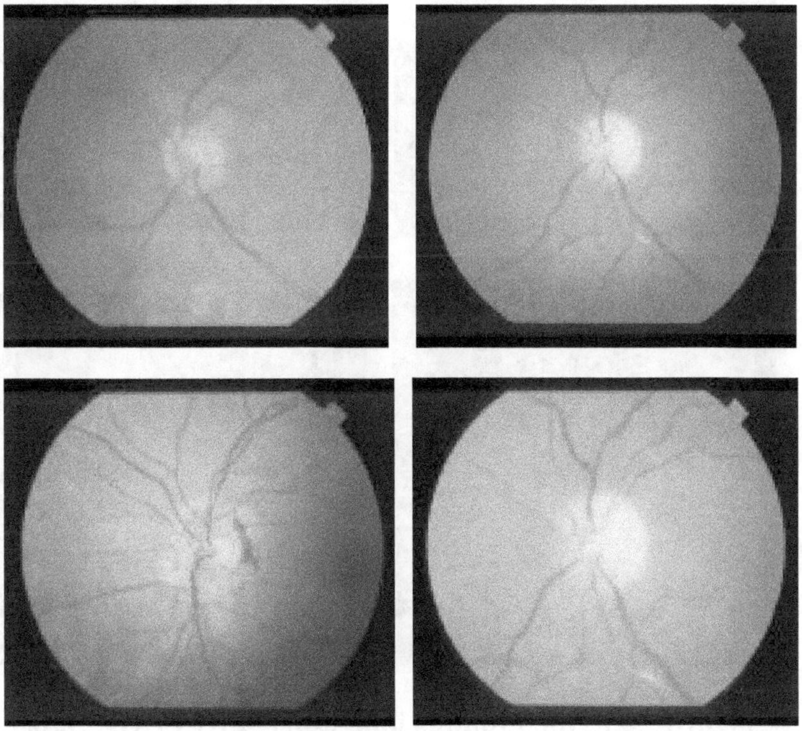

Figure 3.3 Retinal fundus images from STARE Dataset

The REFUGE dataset contains 1,200 retinal images and it is publicly available. The three divisions of this dataset are training, validation and testing. There is an equal amount of images, ie., each division has 400 images. The retinal fundus images are taken with two separate fundus cameras (the Ziess Visucam 500 and the Canon CR-2). CanonCR-2 is utilized to collect the retinal images for validation and testing while Ziess Visucam 500 has been used to capture the images for training. Zeiss Visucam 500 has the resolution of 1634 x 1634 and Canon CR-2 cameras have the resolution of 2124 x 2056 pixels. The Refuge dataset has 1,080 healthy images and 120 images with glaucoma.

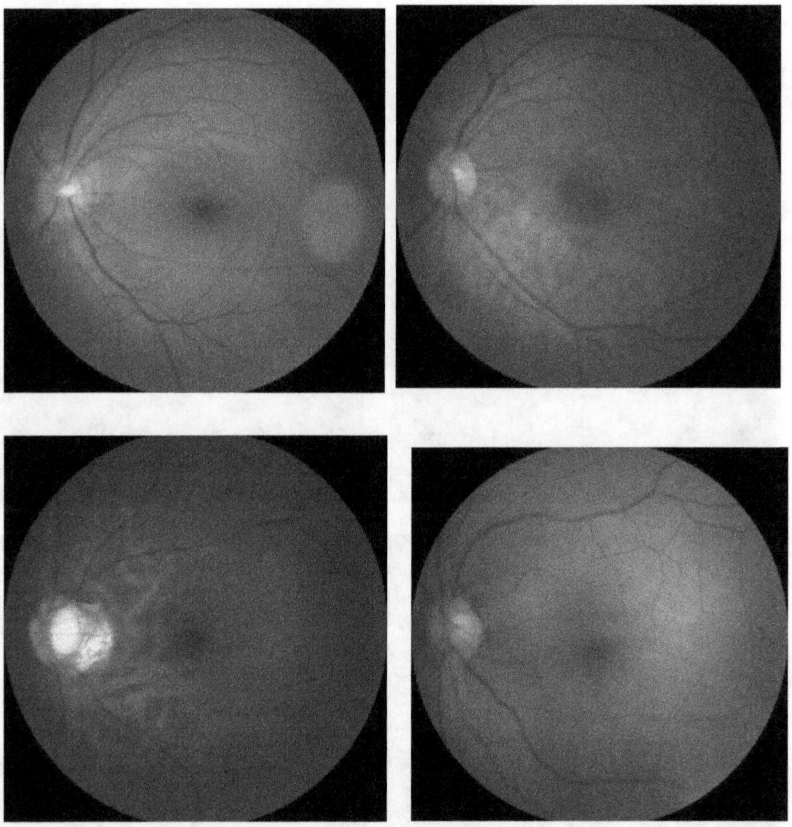

Figure 3.4 Retinal fundus images from REFUGE Dataset

Figure 3.4 shows the retinal fundus images from REFUGE dataset. Table 3.1 shows details of the dataset employed in the current research. In the research, 70% of data are utilized for training and 30% of data are utilized for testing processes.

Table 3.1 Details of Glaucoma Dataset

Dataset	Glaucoma	Non-Glaucoma
ORIGA	168	482
STARE	50	31
REFUGE	120	1080

3.2 IMAGE PREPROCESSING

The important stage in computer vision is preprocessing, which involves removing noise and improving visual data. Preprocessing is used to improve image quality by removing unnecessary distortions and improving visual features that are significant for subsequent processing. It is necessary to reduce the workload involved in image processing. It aims at increasing the classification performance. It is a vital step employed for processing the medical images.

Eye movement, blurness, non-alignment of the camera, retinal artifacts, challenges in camera focus and noise involved in the image collection stages can all affect the quality of a fundus image. Hence, preprocessing the retinal fundus image is necessary. In the current study, it is initiated to use Gaussian filters and histogram equalization techniques for image preprocessing. The Gaussian filter is employed to eliminate the Gaussian noise in the retinal fundus images. Histogram equalization method is then applied to enhance the contrast. After employing the histogram equalization, the pixels can be seen to get spread over the gray levels. Histogram equalization can be used for improving the image quality. The aim of this chapter is to describe the preprocessing of the retinal fundus images for the detection of glaucoma.

3.3 IMAGE PREPROCESSING TECHNIQUES

The quality of an image can be degraded by a variety of circumstances. To use the data effectively, a variety of image processing techniques can be used. The following are a few image preprocessing methods that are used to improve the image quality.

- Pixel intensity transformations such as contrast enhancement, image brightness enhancement and so on.
- Rotation, scaling and other geometric operations.
- Image restoration employing denoising techniques such as Median filters, Gaussian filters, Weiner filters and so on.

3.3.1 Pixel Intensity Transformations

Contrast is the difference between an image's maximum and least pixel intensity. This type of preprocessing is commonly used to increase the contrast in the image. The intensity difference between the target and the background can be raised to increase contrast. The increase in the contrast makes the features in the image more apparent. Further, the brightness of the image can be enhanced by raising its pixel intensity. Histogram equalization, CLAHE, Mean, Median, Gaussian and Gabor Variance Filters can be used to improve the quality of the image.

3.3.2 Geometric Operations

Geometric transformations are commonly utilized in image analysis. By using geometric transforms, the geometric distortion that occurs during the acquisition of an image can be removed. Rotation, scaling, translation and shear are the fundamental transformations. Image rotation is a geometric transformation that rotates an image by a certain angle, like 90 degrees or 180 degrees. Scaling refers to the modification of the image resolution. Translate operator applies a geometric transformation that maps the position of every image element in an input image to a new location in an output image. When an image is sheared, the pixel values are either horizontally or vertically shifted. These transformations are employed to standardize the quality of an image, so that subsequent processing may be applied efficiently.

3.3.3 Denoising Techniques

More precise images are required as a result of an increase in the amount of digital images being captured everyday. However, noise always degrades the images taken, resulting in a decline in the image quality. As a result, efforts are needed to minimize the noise without affecting the image quality. In the medical image processing, noise can be removed during preprocessing for the accurate identification of information from the image. Numerous strategies have been offered by researchers for reducing noise. Filters that are commonly employed to denoise the data are mean filters, median filters, gaussian filters, weiner filters, bilateral filters and so on.

3.4 PROPOSED PREPROCESSING TECHNIQUE

The crucial step in medical image analysis is preprocessing which involves removing noise, contrast enhancement, and thus improving the quality of an image. In the current research, retinal fundus images are preprocessed using Gaussian filter and histogram equalization technique. During the image acquisition process, fundus images are affected by low illumination. Histogram equalization addresses this non uniform illumination as it has a tendency to increase contrast. The core premise is that features are present at lower spatial frequencies, noise occurs at higher spatial frequencies. Gaussian filter is used to reduce high frequency noise in the image. This helps to improve the image quality and to achieve effective classification accuracy.

3.4.1 Gaussian Filter

Generally, the fundus images are disrupted by the Gaussian white noise. An essential component of a disease diagnosis is the elimination of noise from these images. In fundus images, the Gaussian filter is employed to eliminate the Gaussian noise. In the current research, the retinal fundus images are initially preprocessed using Gaussian filter to remove the Gaussian noise in the images. The Gaussian filter is defined by Equation (3.1).

$$g(a, b) = \frac{1}{2\pi\sigma^2} e^{-\frac{a^2+b^2}{2\sigma^2}} \tag{3.1}$$

where a and b represent the value of the coordinates. Sigma (σ) or variance values have an inverse relationship to filtering; lesser sigma values result in more frequencies being suppressed and vice versa. The Gaussian filter is applied to the retinal fundus image to reduce the Gaussian noise. Figure 3.5 shows the result of applying Gaussian filter to ORIGA, STARE and REFUGE datasets. The image is given to the Gaussian filter, and thus, Gaussian noise has been reduced. The outputs from the Gaussian filter are then given to histogram equalization to enhance the contrast of the image.

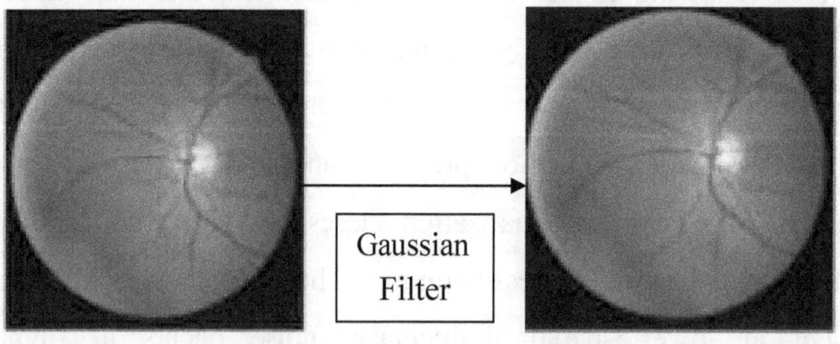

a) Result of applying Gaussian filter in ORIGA dataset

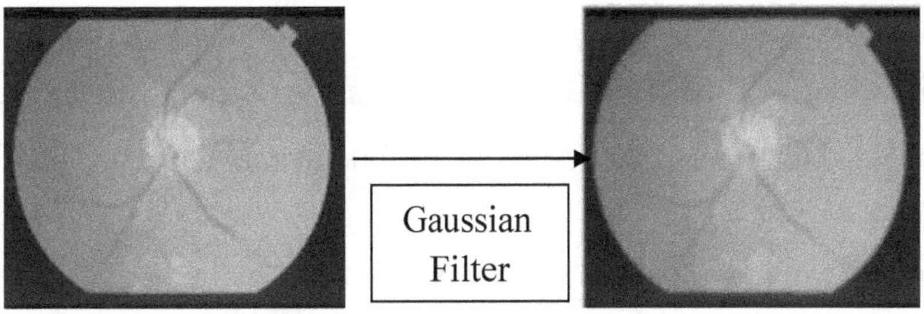

b) Result of applying Gaussian filter in STARE dataset

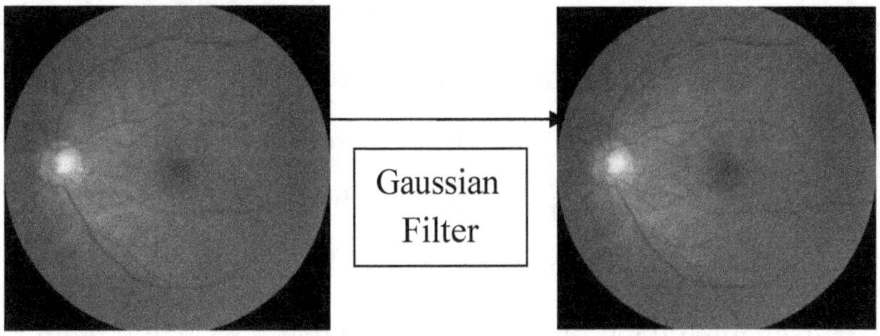

c) Result of applying Gaussian filter in REFUGE dataset

Figure 3.5 Result of applying Gaussian filter a) ORIGA, b) STARE and c) REFUGE datasets

3.4.2 Histogram Equalization

The distribution of numerical information can be typically represented by a histogram. The histogram is the process of showing the occurrence of the intensity of an image. On the other hand, it can also be defined as a graph showing the image pixels at each intensity value of the image.

Histogram equalization is the most widely used technique for contrast enhancement in many applications due to its simplicity and effectiveness. This method can be used for improving the image quality. The fundus image consists of three main channels such as red, blue and green. In general, the fundus images have poor contrast. Histograms can generally be equalized to enhance the image quality. In the current research, histogram equalization has been used for preprocessing to improve the quality of the image and for the accurate prediction of glaucoma.

General working of Histogram equalization

Histogram equalization is one of the image processing approaches to enhance contrast in images. It achieves this by evenly distributing the intensity values. Thus, it makes it possible for regions with less contrast to acquire more contrast. Overall, this approach would be applied either on a part or on a full image to get a uniform histogram. By applying this method, the distribution level of intensity would be changed. If there are peaks and valleys available on the image, it is found that they will remain the same even after the process of histogram equalization, but they both will be shifted. Hence this is called flattening in histogram equalization where each pixel would be given the new value.

In general, the histogram would be applied in three steps:

1. The first level is the formation of the histogram.

2. The second step is to give new values to each intensity value.

3. The third step is to replace the previous intensity of images with the new values.

A histogram is equalized with the initial step of obtaining the histogram for the original image. In the case of a grayscale image with 256 different intensity levels for i, (where i = 0 [black] and 1, 2, 3, 255 [white]), the likelihood that a randomly selected pixel might have the intensity level i as

$$h(i) = p_i \tag{3.2}$$

It is then necessary to get the Cumulative Distribution Function (CDF). The probability H that a randomly chosen pixel would take one of the intensity values between 0 and j makes up the cumulative distribution function H(j). Hence the normalized histogram h(i) that has been arrived from previous step, can be employed for the computation of CDF.

$$CDF = H(j) = \sum_{i=0}^{j} h(i) \tag{3.3}$$

where j = 0,1,2,3,.....252,253,254,255. The normalized histogram has a total of one as its component parts. Hence H(255)=1. Further, it is necessary to calculate the transformation in order to convert the old pixel intensity into new intensity values. Let K be the entire number of intensity values that are feasible (e.g. 256). T(j) in the equation (3.4) represents the new intensity value and j represents the old intensity value.

$$T(j) = \text{floor}((K - 1) * CDF_j \tag{3.4}$$

Finally, a new image is created by substituting the new gray values for the original gray values. As a result, a new histogram that relates to a new output image is produced.

Histogram equalization algorithm

Input: Normal and glaucoma Image

Output: Histogram Equalized image

Step 1: Load the image

Step 2: Acquire the histogram for the image

Step 3: Obtain the CDF

Step 4: Perform histogram equalization on the input images

Step 5: Create a new image by substituting the new gray values for the original gray values.

Step 6: Generate the histogram equalized image and its histogram.

Figure 3.6 shows the original image, with its histogram, and equalized image with the equalized histogram. Once the histogram equalization is done, the pixels can be seen to get spread over the gray levels. The quality of the images is greatly increased by using histogram equalization preprocessing technique.

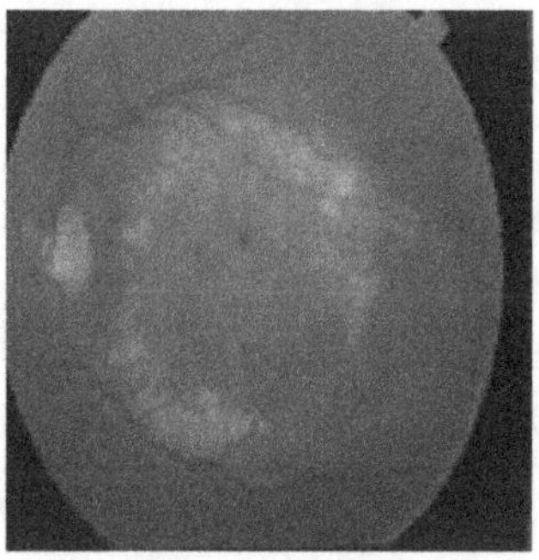

a) Original Image

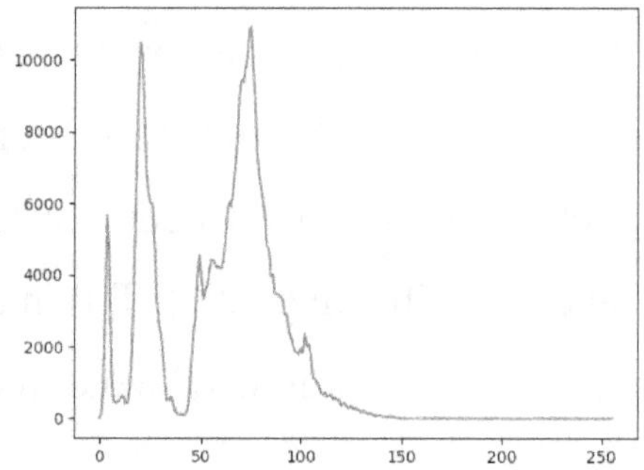

b) Original histogram

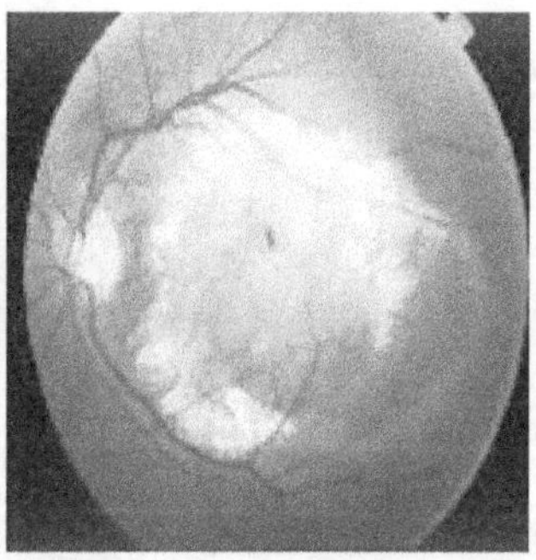

c) Histogram equalized Image

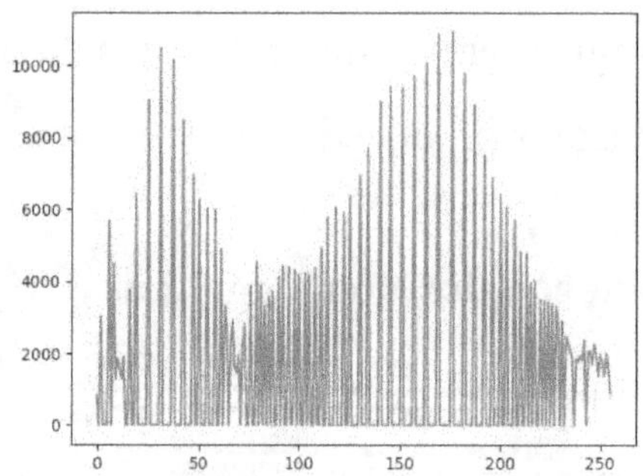

d) Equalized histogram

Figure 3.6 Histogram for original and equalized images

3.4.2.1 Histogram equalization for ORIGA dataset

Figure 3.7 represents the original images and histogram equalized images for ORIGA dataset.

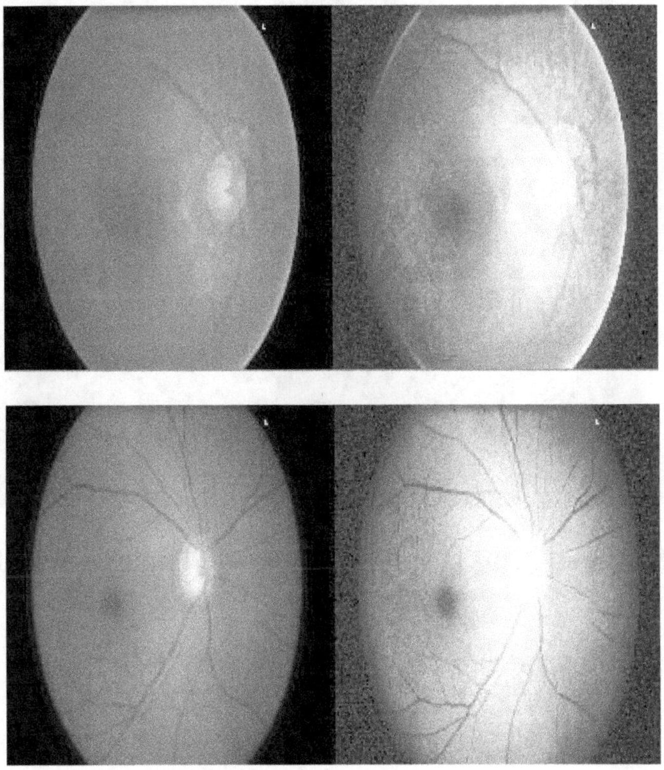

Figure 3.7 Original and Histogram equalized images for ORIGA dataset

3.4.2.2 Histogram equalization for STARE dataset

Figure 3.8 represents the original images and histogram equalized images for STARE dataset.

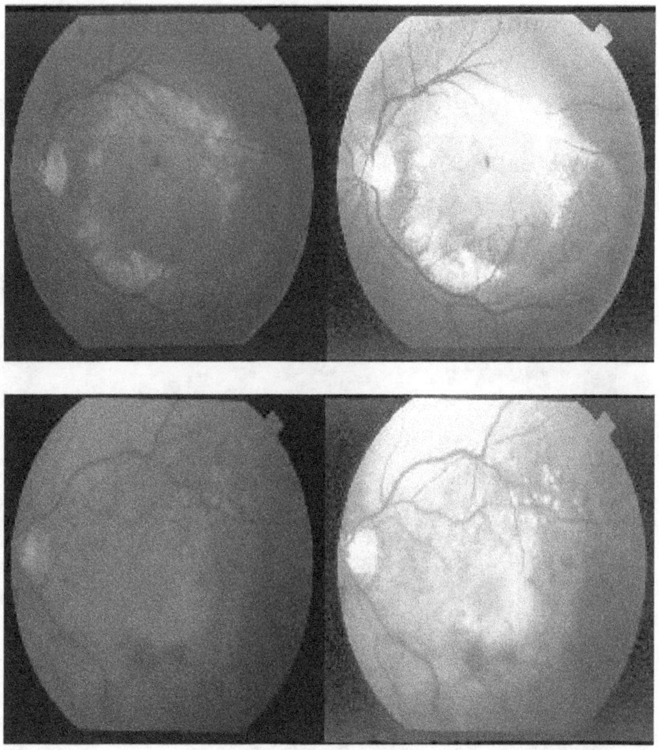

Figure 3.8 Original and Histogram equalized images for STARE dataset

3.4.2.3 Histogram equalization for REFUGE dataset

Figure 3.9 represents the original images and histogram equalized images for REFUGE dataset.

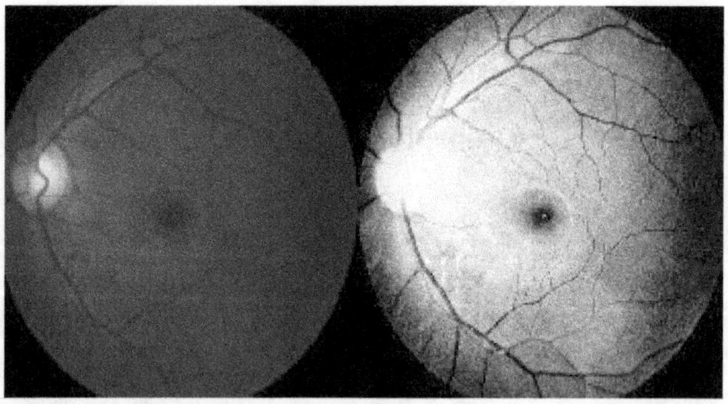

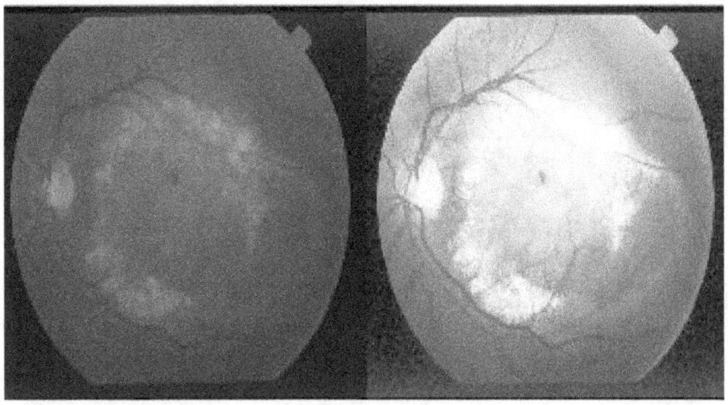

Figure 3.9 Original and Histogram equalized images for REFUGE dataset

3.5 SUMMARY

The fundus images are disrupted by the Gaussian white noise. An essential component of the disease diagnosis is the elimination of noise from these images. Hence the Gaussian filter is employed on the fundus images to eliminate the Gaussian noise. In the current research, the retinal fundus images are initially preprocessed using Gaussian filter to remove the Gaussian noise from the images. Later, Histogram equalization, an effective image enhancing technique has been employed to improve the image contrast. The equalized image shows the complete dynamic range of histogram. Thus, histogram equalization method aims to combine the nearby gray levels together in order to force the number of pixels in each gray level to be uniform. It assigns the intensity values of pixels in the input image to create a homogeneous distribution of intensities in the output image. It boosts contrast, and thus it helps to predict the abnormalities in the image in an easy way.

CHAPTER 4

SEGMENTATION AND FEATURE EXTRACTION METHODS FOR DETECTING GLAUCOMA

Medical image analysis requires two significant phases, namely segmentation and feature extraction. Segmentation is an important strategy in medical imaging. Medical professionals who manually segment images need to trace the patterns slice by slice using some pointing devices like a mouse. This process is drawn out and the results could be tedious due to intra-or-inter viewer variance. Feature extraction is vital for classifying the data and it gives better interpretation. Without feature extraction, the accuracy of detection cannot be effectively achieved. Numerous techniques have been proposed for these phases by various researchers.

4.1 SEGMENTATION

Segmentation is a process of extracting the ROI from an image. Manual segmentation requires an expert personnel and it is subjective. Automated segmentation approaches have been developed to address the problems of manual segmentation. Since automated segmentation methods are typically quicker than the manual segmentation methods and need little user intervention during the segmentation process, these methods have proven to be a viable solution for a variety of applications. In the current study, semantic segmentation approach is used to segment the images.

After segmentation, relevant features are extracted from the retinal images using density correlation method. It has been then examined through the PCA approach for optimal feature selection. Several techniques have been proposed for the computer assisted image segmentation.

4.2 IMAGE SEGMENTATION METHODS

Various segmentation methods have been designed by various researchers. Figure 4.1 shows the different segmentation methods categorized into obsolete, ancient and recent methods (Chowdhary *et al.* 2020).

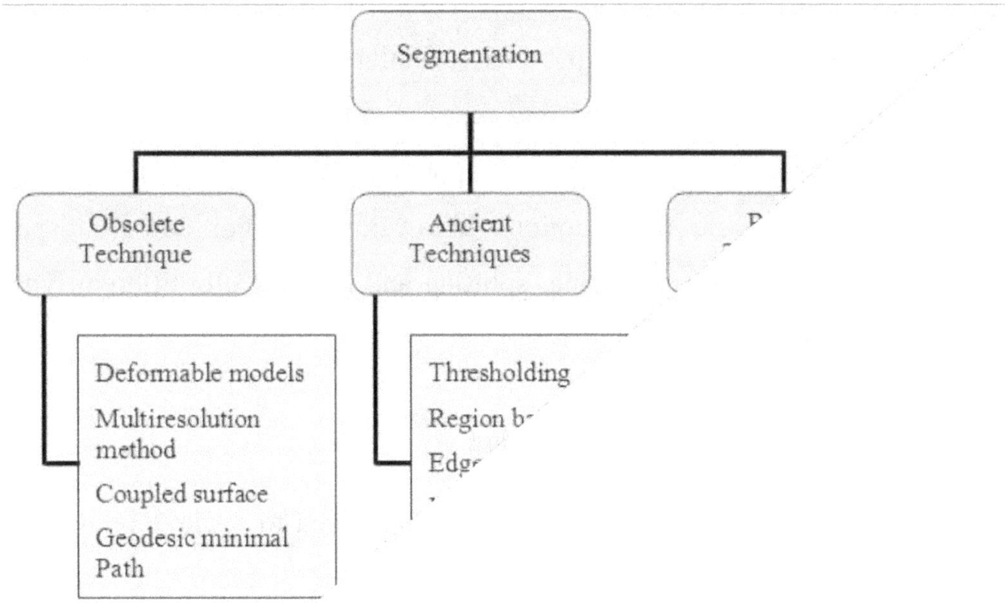

Figure 4.1 Various Segmentation methods

4.2.1 Deformable Models

Deformable models have been applied in medical image analysis (McInerney & Terzopoulos 1996). Many anatomic structures, ranging in size from the macroscopic to the microscopic, have been segmented, visualized, tracked and quantified using 2D and 3D deformable models. Deformable models have been used to locate specific areas in the data. It captures the intricate details and a wide range of shapes found in biological systems.

4.2.2 Region Based Segmentation

In the region based segmentation technique, the nearby pixels are compared for similarity. In this technique, the pixels with similar characteristics are categorized into separate zones. The most popular way to determine similarity in segmentation approaches is to use gray level intensity. However, there are other options like variance, color and multispectral characteristics. Region growing, splitting and merging are different types of region based segmentation.

4.2.3 Atlas Based Segmentation

Atlas is a common technique employed to retrieve relevant prior knowledge from the input images. An atlas can be used in medical imaging to estimate the position of objects (Cuadra *et al.* 2015). This spatial information enables the separation of the objects of interest from the other objects with the same features. Additionally, an atlas can provide details about the texture and structure of the objects. This information enables the ROI to detect the shape of the object. The atlas can also highlight the features of the nearby objects and this makes it easier to separate the objects from the neighborhood.

4.2.4 Active Shape Model

The biomedical images typically have complex structures and these structures would frequently appear differently from image to image. It is difficult to locate certain structures in these images. However, it is possible to examine challenging images successfully by employing the models that can handle the variability. Model based approaches use a prediction from the previous model. Further, it looks for the best fit between the model and the data in new images. Active shape models can be employed to analyze new images by determining the parameters that best match a model instance to an image (Cootes *et al.* 1995). It is a model based technique that uses the predicted features of the image from the previous model and normally seeks to locate the optimal match with the model and the data in new images. It has been effectively used in many applications.

4.2.5 Segmentation using Artificial Intelligence

Image segmentation divides the images into various components. It is a key phase in computer vision. The application of artificial intelligence has helped the computer vision field to grow significantly in recent years. The image segmentation is used in various fields including medical imaging, autonomous vehicles and robotics. In medical imaging, image segmentation has emerged as a vital procedure for the identification of specific anatomical and other regions. The use of artificial intelligence in medical imaging can assist the clinicians in their decision making when working on diagnosis.

4.3 PROPOSED SEGMENTATION APPROACH

Segmentation creates a collection of segments that describe the full object or a group of image contours. To detect points, borderlines and image curves, the fundus images are segmented. The goal of the segmentation is to disintegrate the feature vector into optical disc. In semantic segmentation, every image pixel is classified, allowing objects to be detected and defined. In the current research, U-Net is employed for segmenting the optic disc and the optic cup. Many researchers have employed U-Net for semantic segmentation for their research. Figure 4.2 shows the input retinal fundus image used for experimentation and the outcome obtained from segmentation.

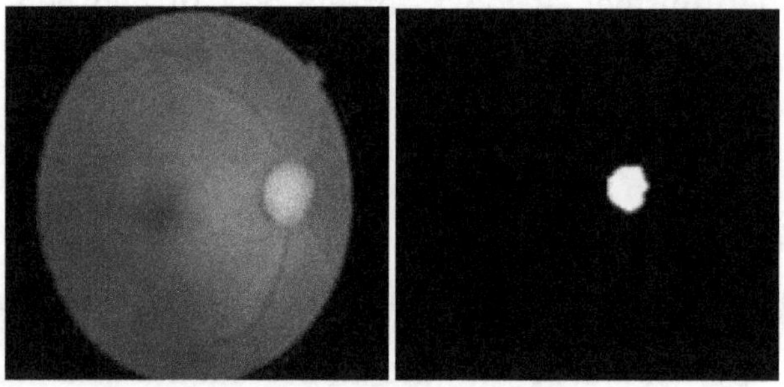

Figure 4.2 (a) Fundus image and (b) segmented outcome

U-Net is a convolutional network employed for biomedical image segmentation. It is considered to be one of the best approaches for semantic segmentation. The semantic segmentation has been diagrammatically represented in Figure 4.3.

The preprocessed images from Gaussian filter and histogram equalization are given as an input to the semantic segmentation. U-Net is used for segmenting the optic disc and the cup from the fundus images. The segmented maps are then fed to the feature extractor and finally to the classifiers.

Figure 4.4 shows the architecture of U-Net (Punn & Agarwal 2022). In the field of biomedical image analysis, U-Net is the best segmentation model since it can incorporate high and low level information, allowing the extraction of complicated features and increasing accuracy. The U-Net architecture consists of a contracting path and an expansive path.

Feature extraction is carried out in the contracting path. The contracting path has a sequence of convolutional layers, followed by max pooling and ReLU layers. However, the expansive path integrates the spatial data with the features of high resolution to create the necessary segmentation map.

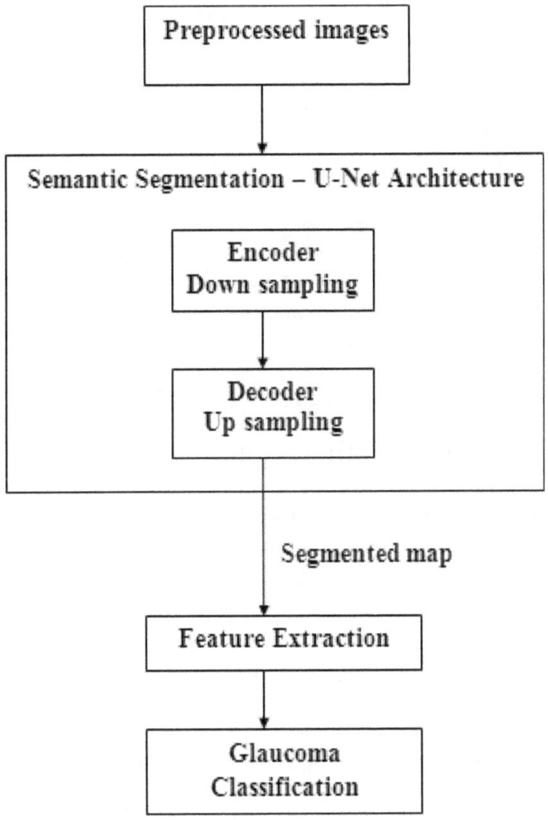

Figure 4.3 Diagrammatic representation of Semantic Segmentation

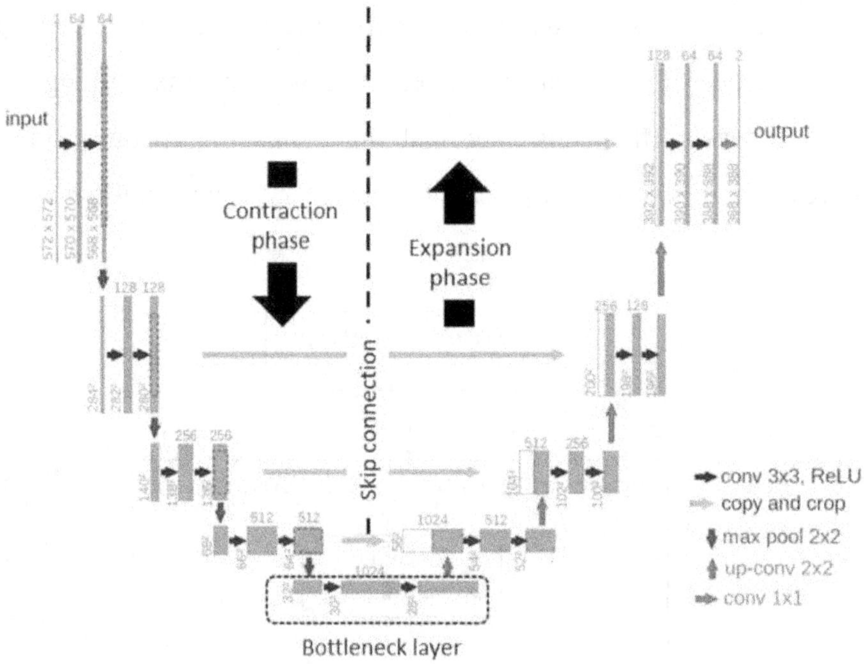

Figure 4.4 Architecture of U-Net

The input image is first put through a convolutional layer. Then the activation function ReLu has been applied using Equation (4.1) given below

$$ReLu(a) = \max(0, a) \tag{4.1}$$

where a is the input given to the neuron.

Furthermore, 2X2 max pooling is applied to lower the dimensions of the input layer by choosing the peak value from every cluster of neurons and transforming into a single neuron in the subsequent layer. The model optimization includes the accurate adjustment of parameters including weight initialization. The inputs are normalized to the network layers using batch normalization with the statistical information like mean and covariance. With batch normalization applied to each layer in the U-Net model, the overall layer activation has been lowered to zero. The dropout layer, which is an aspect of regularization, is used to address the issue of overfitting. Moreover, Adam optimizer is utilized and binary cross-entropy is used for loss function.

Finally, the output values of the segmentation map are limited between 0 and 1 by the Sigmoid function represented in Equation (4.2).

$$S(a) = \frac{1}{1 + e^{-a}} \tag{4.2}$$

The performance measures used for the evaluation of segmentation in the current research are IoU and Dice score. These IoU and Dice are the common evaluation metrics for image segmentation and stated in Equations (4.3) and (4.4) respectively. IoU can also be referred to as Jaccard Index. These metrics gives a prompt assessment of segmentation from the U-Net model.

$$IoU = \frac{|A \cap B|}{|A \cup B|} \tag{4.3}$$

$$Dice = \frac{2*|A \cap B|}{|A| \cup |B|} \tag{4.4}$$

where A represents the predicted map and B represents the accurate binary map. The segmented images are then given to the feature extractor to extract the essential features needed for the classification of glaucoma. U-Net appears to be a quite effective segmentation method for medical images.

4.4 FEATURE EXTRACTION

One of the most key design considerations for a classification problem is choosing an appropriate feature set (Davies 2004). The primary objective of a feature extraction is to extract the most pertinent information from the image. When the input is extensive to process or when there is a possibility that it is redundant, the data can be turned into a smaller set of features (Kumar & Bhatia 2014). The process of extracting features from the input data is called the feature extraction. Extracting the features from an image is essential as machine learning cannot handle the immense amount of image data. Due to the large number of variables used, the performance evaluation of repetitive data is a massive issue. Feature extraction minimizes the dimensionality of the image representation. It is used to achieve sufficient accuracy for the real time applications. Many machine learning specialists consider that one of the essential components for effective model construction is the feature extraction. In the current research, the density with correlation based feature extraction approach and PCA are used for extracting the features.

4.5 FEATURE EXTRACTION METHODS

Many researchers have used various feature extraction methods like Local Binary Pattern (LBP), Gray Level Co-occurrence Matrix (GLCM), Gabor feature extraction, Independent Component Analysis (ICA), Principal Component Analysis (PCA) and so on.

4.5.1 Local Binary Pattern (LBP)

Local Binary Pattern (LBP), a texture descriptor, is used to recognize textures and patterns in the images. It must be computed for every pixel of the image. It assigns a binary number to every pixel present in an image by analyzing its gray level to that of its nearby pixels. It has been employed in various applications. It can be used to summarize texture pattern probability into a histogram. The distribution pattern of LBP histogram can be used to determine the regularity of the texture. It uses the Euclidean distance method to categorize the texture pattern that has been produced from LBP computation. The results of using LBP indicate that the texture features retrieved can be employed as an input for pattern categorization.

4.5.2 Gray Level Co-occurrence Matrix (GLCM)

Gray Level Co-occurrence Matrix (GLCM) also referred to as the gray level spatial dependence matrix, is a statistical technique utilized for analyzing texture which considers the spatial relationship of pixels. Vertical, horizontal and diagonal are the three orientations of analysis for GLCM. The GLCM is produced from a grayscale image. To identify the texture of an image, the GLCM functions determine the occurrence of the pairs of pixels with particular values in a specific spatial relationship, construct GLCM and then retrieve the statistical measures from this matrix. A few of the statistical measures that can be obtained from GLCMs are correlation, energy, contrast, homogeneity and entropy. These statistics give information about the texture of the image.

4.5.3 Gabor Feature Extraction

Gabor feature extraction is one of the feature extraction methods utilized for extracting the gabor features from the data. In image processing, gabor features have been widely used. In Gabor feature extraction, it is essential to apply Gabor filter and then dynamically regulate the output of the Gabor filter to improve the performance in the presence of uneven lighting and background noise. Later the histogram features are extracted and finally, more relevant specific features are obtained.

4.5.4 Independent Component Analysis (ICA)

Independent Component Analysis (ICA) is a method for extracting features from the data and it is utilized in many applications. The distinctive feature of ICA is its ability to extract statistically independent components from the input data. This method is considered as unsupervised learning as it produces a set of vectors with the most independent components possible.

This unsupervised technique, by nature, is tied to the input vectors and it cannot guarantee high performance in classification applications.

4.6 PROPOSED FEATURE EXTRACTION TECHNIQUES

The removal of unnecessary data, reduction of dimensionality and improvement of learning accuracy can all be achieved by feature selection (Gopika *et al.* 2018). In the present research, subsequent to the segmentation phase, the features could be chosen using PCA and correlation based approach.

4.6.1 Correlation Based Feature Extraction

Correlation based feature selection combines suitable correlation measure with a heuristic search method. It considers the correlations among different features when evaluating feature subsets. Hence, in order to search the relevant features, this method computes feature-class and feature-feature correlations.

Julesz (1981) has concluded that the first and second-order texture characteristics appear to be significant in focusing the attention towards particular textured regions based on various researches. The first-order statistic features give details about the distribution of the gray levels of the images. The first-order statistics such as mean, variance and skewness and the second-order statistics such as correlation, contrast, homogeneity and entropy are used to represent the texture of an image. First and second-order statistics provide a fewer distinct features needed for the classification of images.

The density based correlation approach retrieves features of second-order statistical textures. It has been utilized in a variety of applications. It is employed in third order and higher-order textures with 3 or more pixels. It evaluates the subset based on the prediction and redundancies of every function. The heuristic meaning of subset differentiates the density based correlation method from the other methods. This indicates that if a heuristic function is given, then the algorithm can select the option that maximizes the result of this function in each of its subsequent steps. The key features like entropy, energy, homogeneity, contrast and correlation can be obtained using the equations given below. The entropy calculates the distribution of the gray levels. It is obtained using Equation (4.5).

$$Entropy = \sum_{i,j=0}^{n-1} -\ln(p_{ij})p_{ij} \tag{4.5}$$

Additionally, energy and homogeneity are defined in accordance with Equations (4.6) and (4.7) respectively.

$$Energy = \sum_{i,j=0}^{n-1} (p_{ij})^2 \tag{4.6}$$

$$Homogeneity = \sum_{i,j=0}^{n-1} \frac{p_{ij}}{1+(i-j)^2} \tag{4.7}$$

Contrast provides a measure of the intensity of a pixel with its neighbor and can be computed using Equation (4.8)

$$Contrast = \sum_{i,j=0}^{n-1} p_{ij}(i-j)^2 \tag{4.8}$$

Correlation gives the details of association of the pixels among their neighbors. The correlation across all the image pixels is calculated using Equation (4.9).

$$Correlation = \sum_{i,j=0}^{n-1} p_{ij} \frac{(i-\mu)(j-\mu)}{\sigma^2} \tag{4.9}$$

where n represents gray level count, μ represents mean, σ represents variance and p_{ij} represents the i_{th} and j_{th} element of correlation matrix. It is better to ignore irrelevant features as they have less correlation with the class. Redundant features must be eliminated as they would be closely connected with the other or more of the remaining features. Pearson correlation coefficient is employed to obtain the correlation. The acceptability of a feature depends on the levels it detects the classes. The feature subsets in correlation based feature extraction are evaluated using Equation (4.10)

$$M_X = \frac{n\overline{r_{cf}}}{\sqrt{n + n(n-1)\overline{r_{ff}}}} \tag{4.10}$$

where r_{cf} stands for the mean correlation between feature-class, r_{ff} stands for the average feature-feature intercorrelation, n represents the number of features and M_X is the heuristic "merit" of a feature subset x having n features.

The correlation between two features is measured using the Pearson correlation coefficient. In the current research, two features are defined as very correlated if their Pearson correlation coefficients have an absolute value larger than 0.9. The evaluation focuses entirely on the correlation of feature-class for the initial iteration. The empty subset acquires the feature that has a maximum feature-class correlation. Furthermore, in the subsequent steps, every feature, excluding the one that has already been added, is assessed and the best subset is retained. This process is continued till the algorithm examines the entire subset of features. The feature subset with the greatest merit is returned by the algorithm. In the current research, the subset with the greatest merit that has 45 features has been obtained.

The steps involved in the correlation based feature extraction method are listed below.

Algorithm : Correlation based Feature Extraction

Input: Dataset containing n features

Output: Extracted features from correlation method

Step1: Load the segmented data

Step 2: Calculate feature correlations: feature-class and feature-feature

Step3: Compute the merit (evaluation function)

Step4: Return the feature subset with the greatest merit

The elements in the correlation based feature extraction have been depicted in Figure 4.5. The preprocessed image is given as the input to the segmentation algorithm. The output of the segmented image is given to the correlation based feature extraction algorithm as the input. The correlation between feature-class and feature-feature is computed. The feature subset with the greatest merit is returned by the algorithm. PCA is used to select the optimal features. The output from the feature extraction process is given to the machine learning algorithms, namely neural network, SVM and random forest to classify the images into either normal or glaucoma.

Thus, the current research intends to extract features from the results of segmentation obtained already. The clinical features like CDR, Neural Retinal Rim (NRR) and a few texture features are extracted from the segmented output for glaucoma diagnosis. The texture features like mean, standard deviation, correlation, entropy, contrast, energy, sum entropy, skewness, homogenity, kurtosis, sum average, sum variance, difference variance, difference entropy, autocorrelation, inverse difference and maximal correlation coefficient are extracted. The correlation based feature extraction method identifies the feature subset having low feature-feature correlation in order to prevent redundancy and high feature-class correlation in order to improve the performance.

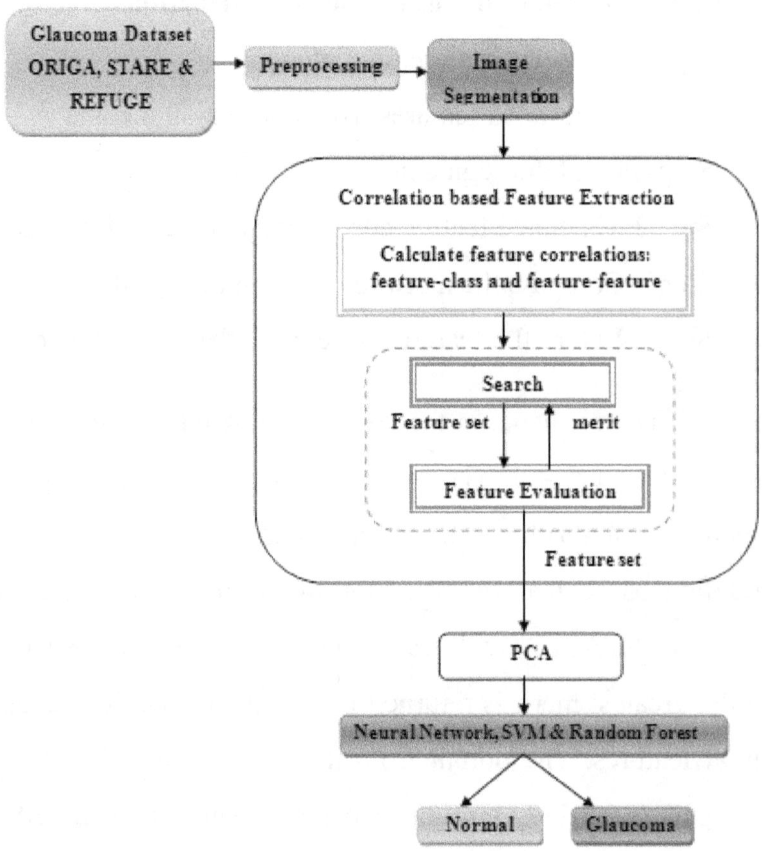

Figure 4.5 Diagrammatic representation of Correlation based Feature Extraction

Extraction of CDR

Clinical authorities use CDR value to detect the progress of eye diseases. CDR should be less than 0.5. Glaucoma can be inferred from a high cup to disc ratio. The most popular feature for glaucoma disease detection is the CDR. The pattern of cupping happens whenever a person experiences glaucoma as a result of a significant amount of strain placed on the retina. The optic cup and the optic disc area are employed to calculate CDR using Equation (4.11).

$$\text{CDR} = \frac{\text{Area of Optic Cup}}{\text{Area of Optic Disc}} \tag{4.11}$$

A CDR value of less than 0.5 is regarded as healthy, but more than that indicates the presence of glaucoma (Ahmad *et al.* 2014).

Extraction of ISNT

The shrinking of NRR area is one of the key features utilized to diagnose glaucoma. To diagnose glaucoma, a classification process can use the ISNT rule verification feature. NRR area is the space between the optic disc and the optic cup. The two techniques utilized to confirm the ISNT rule in NRR region are given below.

i) As per the ISNT rule, the NRR region for a normal eye must be the highest to the lowest of thickness as given in Equation (4.12).

$$I > S > N > T \tag{4.12}$$

ii) The ISNT rule can also be validated using the NRR area ratio. The ratio between the sum of the area in the inferior and the superior region to that of the nasal and the temporal regions represents the NRR. The mathematical representation for NRR area is given by Equation (4.13).

$$\text{NRR} = \frac{\text{Sum of area in Inferior and Superior region}}{\text{Sum of area in Nasal and Temporal region}} \tag{4.13}$$

The NRR area for a healthy eye is more than one, whereas the NRR area for a glaucoma-affected eye is close to or less than one (Kausu *et al.* 2018). The unnecessary features are removed because they do not provide much contribution to the classification process; indeed, they decrease the computing speed and even the classification accuracy. The relevant information is retrieved from the segmented fundus images using clinical features and texture features. The best sets of features are chosen from the retrieved features using a feature selection process to prevent a few unnecessary features for glaucoma classification. In the current research, correlation based feature selection is applied to extract 45 features. Another feature extraction algorithm called PCA is also employed to choose the optimal features.

4.6.2 Principal Component Analysis (PCA)

Feature extraction effectively extracts certain features from the dataset. In machine learning, feature selection is frequently employed as a preprocessing step. The process of feature selection involves the locating of the unnecessary features, eliminating the redundant features from the dataset and deleting the untrustworthy data. PCA is used to select the most optimal features. After correlation based feature extraction, PCA is employed to select the optimal features for glaucoma classification.

Working of PCA

PCA is one of the effective tools employed for data analysis and its primary use is dimensionality reduction. The features that are retrieved are processed using PCA to select the optimal features. In PCA, to reduce the dimensionality of the data, a covariance analysis between variables is made. The primary objective of PCA is to compute the principle components. To find the principal components, it is essential to compute the covariance matrix, eigenvectors and eigenvalues of the covariance matrix. Further, it is essential to sort the eigen vectors in decreasing order of eigen value (Al-Naser *et al.* 2012). The sorted patterns of eigen vectors serve as the outcome. The steps involved in PCA are given below.

Algorithm: PCA

Input: Extracted Features

Output: Optimal features

Step 1: Subtract mean from every dimension of data.

Step 2: Calculate covariance matrix for analysis.

Step3: Compute eigenvectors and eigenvalues of the covariance matrix.

Step 4: Eigen vector with the most significant eigen value is selected as the principal component feature.

After all the features have been retrieved, the PCA method is, thus, used to choose the optimal features for glaucoma classification. The principal components are selected and a new feature vector formation has been accomplished through PCA. Finally, 9 features are found to be highly relevant for the glaucoma diagnosis using PCA. The suggested system minimizes the complexity and eliminates unnecessary features by using PCA. By reducing the irrelevant features, the accuracy of glaucoma diagnosis is increased. The features chosen from PCA and correlation based method are then used for the classification of glaucoma using the machine learning algorithms. The effective feature selection increases the effectiveness and accuracy of glaucoma classification.

4.7 SUMMARY

Numerous techniques have been proposed for segmentation and feature extraction phases by various researchers. The preprocessed images are given as the input for segmentation process. Semantic segmentation is employed to segment the optic disc and the optic cup. The segmented image serves as the input for the feature extraction algorithms developed. The relevant information is retrieved from the segmented fundus images using clinical features and texture features. Correlation based feature extraction has been used to extract different features from the images. The best sets of features are chosen from the retrieved features using a feature selection process in order to prevent a few unnecessary features for glaucoma classification. In the suggested methodology, a PCA has been used to identify optimal features. The features considered to be highly relevant for the diagnosis of glaucoma are CDR, ISNT ratio of NRR, Contrast, Correlation, Homogeneity, Entropy, Energy, Mean and Variance. It is observed that incorporating structural and textural features for glaucoma classification yields good results. After feature extraction, classification is used to classify the images into either normal or glaucoma. The features discovered in the previous step (feature extraction) form a vector, which serves as the inputs to the classifiers. Neural Network, SVM and Random Forest are the classifiers employed in the current research. The effective feature selection has increased the effectiveness and the accuracy of glaucoma classification.

CHAPTER 5

CLASSIFICATION ALGORITHMS FOR DETECTING GLAUCOMA

5.1 INTRODUCTION

Artificial Intelligence is a field of computer science that deals with the modeling of intelligent behavior in computers. Machine learning is a branch of artificial intelligence. It identifies the patterns in data without requiring human programmers to control every aspect of the classification. However, it involves manual feature extraction. There is no certainty that the human-extracted characteristics are the best ones for use by the classifier because producing the features might be a highly specialized operation requiring significant domain expertise. Moreover, deep learning is a subset of machine learning and it automatically learns features from the data. It has the main advantage of reducing the need for domain expertise. One can input the raw data to a learning algorithm that would automatically learn its most significant features from data, rather than trying to manually select the relevant characteristics from the data.

In the current research, the machine learning algorithms like Neural Network, SVM and Random Forest and the deep learning algorithms such as VGG16, Inception V3 and Inception-ResNet-V2 are employed. This chapter focuses on both the machine learning and deep learning algorithms employed for glaucoma diagnosis.

5.2 GLAUCOMA CLASSIFICATION USING MACHINE LEARNING

The advancement of modern technology has revealed the importance of the diagnosis of glaucoma. Various classification techniques have been developed during the past few decades. Machine learning methods have shown to be crucial for successful research outcomes. Machine learning learns and identifies particular characteristics or patterns in images. In the current research, the three machine learning algorithms, namely Neural Network, SVM and Random Forest are used for glaucoma diagnosis. By acquiring insights from the three supervised algorithms, glaucoma classification model has been developed to create a robust application.

5.2.1 Neural Network

In terms of artificial intelligence, neural networks are the most effective methods of solving practical issues. The structure of a neural network is influenced by the human brain, and it resembles the way biological neurons communicate with one another. A node of an Artificial Neural Network (ANN) consists of an input layer, one or more hidden layers and an output layer. Each node in the network is connected to the other neuron and it has both weight and threshold associated with it. A node gets activated and begins transferring data to the next layer in the network if its output is greater than the predefined threshold value. Otherwise, the data are not transmitted to the next tier in the network.

Neural networks enable computer programs to identify patterns and resolve common issues by modeling the behavior of the human brain. Training data is essential for neural networks to develop and enhance its accuracy. However, if the learning algorithms are fine-tuned, they become effective tools to classify the data efficiently.

The initial perceptrons are shallow neural networks with only one input layer, one output layer and possibly one hidden layer. Deep neural networks are neural networks that have multiple hidden layers. A deep neural network may resolve any challenging real-world problem if there are sufficient numbers of hidden layers of neurons.

In the current research, it is proposed to utilize neural network for glaucoma diagnosis since the neurons learn rapidly and effectively on their own and function well in real-world conditions.

One input layer, two hidden layers, and one output layer make up the multilayer perceptron model, a fully connected deep neural network. ReLU has been used as an activation function for hidden layers while Sigmoid function is employed on the output layer. Binary cross entropy, which is represented in Equation (5.1), is employed as a loss function because it tends to reduce the discrepancy between the actual and the predicted values.

$$BCE = -\sum_{i=1}^{N=2} x_i \log(Sigm(y_i)) \tag{5.1}$$

where x represents the ground truth values, y represents the predicted values, Sigm () represents Sigmoid function, BCE represents Binary Cross-entropy and N represents Number of Classes. As learning rates are calculated depending on the parameters, Adam is utilized as an optimizer to help avoid saddle points while training. The steps involved in the neural network are listed below.

Algorithm: Neural Network

Input: Preprocessed glaucoma dataset

Output: Classify labels

Data split: 70% for training and 30% for testing

Step 1: Load the ORIGA, STARE and REFUGE datasets.

Step 2: Construct Neural Network architecture

Step 3: Train the Neural Network

Step 4: Apply forward propagation

Step 5: Apply back propagation

Step 6: Repeat the process for a specified number of times mentioned in epochs. After training, the neural network successfully figures out weights by minimizing the error.

Step 7: Utilize a network to make a prediction.

5.2.2 Support Vector Machine (SVM)

The typical tasks in machine learning include classifying the data, doing regression analysis and identifying the outliers. SVM is a supervised learning technique employed for the above tasks. It is used frequently for classification. It is chosen as a classifier to find a decision function.

Cortes and Vapnik have designed the SVM algorithm (Cortes & Vapnik 1995). SVM, a popular classification approach, continues to stand out among the most popular robust prediction techniques that may be employed in a variety of classification related applications. To accurately separate two or more different classes, it seeks out an ideal separation line, or hyperplane, i.e.) in n-dimensional classification space, it locates the best hyperplane with the largest margin between the classes.

SVM can deal with linear and non-linear data and can perform both classification and regression. In the case of linearly separable data, the objective of using the SVM technique is to determine the best hyperplane separation. In contrast, with non-linear data, the usual strategy is to use kernel methods to transfer the initial feature space to a higher-dimensional space. Thus, the high-dimensional feature space is created with the use of the non-linear kernel function.

Furthermore, the support vector points are the ones that are closer to the hyperplane and the margins are the distance of vectors from the hyperplane. It is necessary to select a hyperplane that has the largest feasible margin between it and any point in the training set, increasing the likelihood that new data would be correctly classified. A suitable decision boundary is learnt by SVM to maximize the margin. The objective of SVM is to construct a model that can predict the key parameters from the data. In the current research, data are the input image sets that have been altered as a result of applying Gaussian filter, histogram equalization, segmentation and feature extraction methods in the earlier steps.

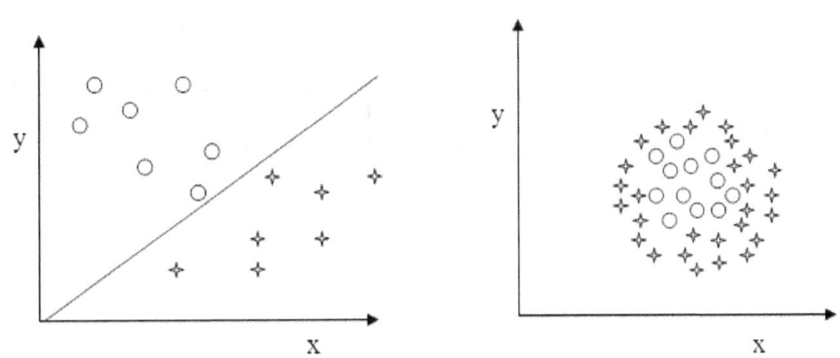

Figure 5.1 Linearly and Non-linearly separable data

Figure 5.1 shows the linearly and non-linearly separable data. The dataset is typically not linearly separable in real time as seen in Figure 5.1.

The SVM linear classification function can be represented by

$$g(x) = W^T X + b \tag{5.2}$$

where g(x) represents the hyperplane, W represents the weight vector, b represents the bias and X represents the data point.

The non-linear decision boundary could be determined using a kernel function that effectively transforms the features into a high-dimensional space. There are various types of kernels employed with SVM. Linear kernel, non-linear kernel, polynomial kernel, Radial Basis Function (RBF) kernel, gaussian kernel are the commonly used ones.

In the current research, SVM is developed to classify the images into two distinct classes either as normal or glaucoma. The SVM classifier, employing RBF kernel, is utilized for classification and it is given in Equation (5.3).

$$K(X_i, X_j) = \exp(-\gamma \| X_i - X_j \|^2) \tag{5.3}$$

The gamma γ parameter specifies the influence range of the training sample while $\|X_i - X_j\|$ represents the Euclidean distance between X_i and X_j. The steps involved in SVM are listed below.

Algorithm: SVM

Input: Preprocessed glaucoma dataset

S=(A,B); A(training images in array), B(Class Label: Normal and Glaucoma)

Output: Classify labels and analyze the performance of the model

Data split: 70% for training and 30% for testing

Step 1: Load the dataset

Step 2: Construct SVM classifier model

Step 3: Train the model

Step 4: Utilize a network to make a prediction

Step5: Analyze the performance of the model.

5.2.3 Random Forest

A popular supervised machine learning algorithm utilized for classification is Random Forest. Random Forest makes use of ensemble learning, a method employed for solving complicated issues by combining a number of classifiers. It consists of a set of decision trees. It integrates the results of various decision trees to get a single outcome. Decision trees are created by random forest on the samples of data that have been chosen at random, predictions are obtained from each tree and the best solution is then chosen by voting. The most popular ensemble techniques available are bagging and boosting.

Bagging is an ensemble approach that applies several models on various training dataset subsets. It then merges the predictions of all the models. If the same input is provided, then the likelihood of obtaining the same results is high. Hence, in this case, bootstrapping is employed. When replacement is performed, the models are provided with distinct data points.

Boosting is a common ensemble training algorithm used to sequentially build the classifiers for random forest that includes independent decision trees. The sequential process of the boosting technique involves each model attempting to solve the flaws of the previous model. Thus, the preceding model serves as a foundation for the succeeding models.

Working of Random Forest

Random Forest algorithm can be used to resolve the classification problems. The key hyperparameters for random forest algorithms that must be set prior to training are node size, total number of trees and total number of sampled features. Random Forest classification uses an ensemble process to achieve the result. It consists of a set of decision trees. Every decision tree in the ensemble composes of images from the training dataset with replacement referred to as bootstrap sample. Further randomness is introduced using feature bagging, increasing dataset diversity and lowering decision tree correlation. In this, for the classification problem, either the normal or the glaucoma class would be determined by a majority vote.

The steps followed in the Random Forest algorithm are given below.

Step 1: Random sample selection from dataset

Step 2: Build a decision tree for the entire samples. This provides the prediction result from every decision trees.

Step 3: For each predicted outcome, accomplish voting

Step 4: Choose the final prediction outcome that receives the most votes.

Pseudocode: Random Forest Classifier

```
Begin:
    Build r classification models
    for a=1 to r do
        Random sampling of the training data G occurs instead of
        constructing the Gi produce a root node, Ni containing Gi
        Run Build_DTree (Ni)
    End for
    Build_DTree (N):
    When N contains the samples of each class, then
        return
    Else
        Randomly select z% of the expected splitted features in N
        Select the feature F with the highest IG for splitting
        Produce c child nodes of N, N1 ……, Nc where C hold c
        probable values
  (C1… ..........., Cc)
    For a=1 to c, do
        Assumption:- Data:   Na to Da where Gi points out all the
        instances in N that match Ca
        Run Build_DTree (Ni)
```

End for

End if

End

Neural Network, a machine learning algorithm, has the potential to be one of the more widely used algorithms. However, SVM and Random Forest are considered as simpler, and more manageable techniques that consistently produce results with high accuracy. In short, glaucoma might be diagnosed with deep learning and as it advances in complexity; it might become a reliable tool for the early diagnosis of glaucoma.

5.3 GLAUCOMA CLASSIFICATION USING DEEP LEARNING

Machine learning algorithms are not very effective at learning features from the raw data, therefore in their most basic form, they still need a human-designed code to turn raw information into input features. Hence there is no assurance that the human-extracted characteristics are the best ones for use by the classifier because producing these features might be a highly specialized operation requiring significant domain expertise. In the current research, images have been classified using straightforward traditional machine learning methods like Neural Network, SVM and Random Forest. However, a sophisticated deep learning method, like CNN, is needed to identify glaucoma. A complicated deep learning model produces a high accuracy than the traditional machine learning methods. This highlights the necessity for a robust algorithm, the deep learning algorithms, to classify the images into either normal or glaucoma.

The CNN belongs to the intense part of neural learning networks and it plays a major role in image recognition and classification processes. Transfer learning concept is being employed in the deep learning models.

Transfer learning focuses on customizing and reusing the features that the deep learning models have learnt on the primary task. Instead of using random generation, transfer learning initializes the parameters based on prior learning. The initial layers intrinsically acquire the primary features like edges, textures and so on and the top layers are more specialized to the application.

In the current research, the aim of deep learning methods is to classify the images either into normal or glaucoma and it is achieved by using the pre-trained convolution algorithms like VGG16, Inception V3 and Inception-ResNet-V2. By optimizing the model parameters, the deep learning algorithms have provided a more precise glaucoma classification model.

5.3.1 Convolutional Neural Network (CNN)

Convolutional Neural Network (CNN), a subclass of deep learning networks, is employed in many deep learning based intensive applications in computer vision. To analyze spatial patterns and carry out tasks like image classification and object detection, CNNs consists of one or more convolutional layers, which are made up of sets of filters. These filters have the ability to automatically extract the features from images, eliminating the need for manual work in choosing features, which is a significant drawback of conventional machine learning algorithms.

Figure 5.2 shows the architecture of CNN. This consists of two major components for extracting the features and classification. Feature extractor includes various modules like multiple convolutional layers, activation functions and pooling layers, whereas the classifier includes a fully connected layer that makes use of the output from the convolutional process to determine the classes of the image using the features that have been previously extracted.

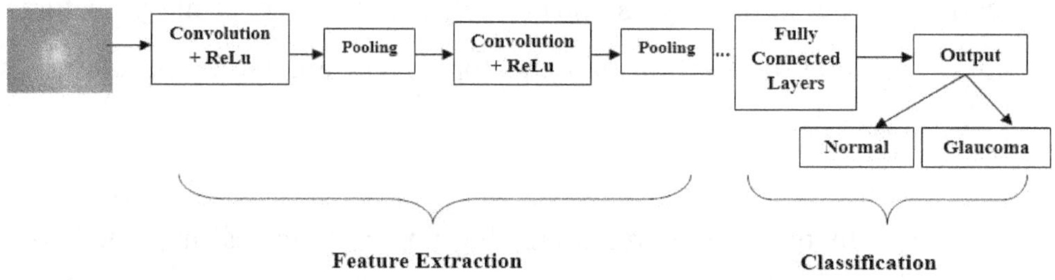

Figure 5.2 Architecture of CNN with different modules

5.3.1.1 Convolution

A major part of CNN is the convolutional layer that consists of many filters (kernels) applied to the input image. The output from the convolutional layer is the feature map which is the representation of the image with filters applied to it. Feature maps provide details about an image like its corners and edges. Several additional features from the input images are learnt by providing the feature map to subsequent layers. Convolutional layers could be layered to build more sophisticated models that can extract further finer features from the images.

5.3.1.2 ReLU layer

ReLU layer serves as an activation function which is one of the most key aspects in the CNN model as it gives non-linearity to the network. All the negative input values are changed to zeros by the ReLU layer. Apart from ReLU, Softmax, tanH and Sigmoid are a few examples of regularly used activation functions. Each of these operations has a specific use. Sigmoid and softmax functions are favored for a CNN model for binary classification and softmax is typically employed for multi-class classification.

5.3.1.3 Pooling layer

The process of pooling can progressively minimize the size of the input while maintaining essential aspects in the images. Max and average pooling are the two main kinds of pooling. The average pooling gets the average value from every feature map, whereas the max pooling gets the maximum value. The parameter count in the model is lowered by the pooling layer. This helps the network to train faster, avoid overfitting and to be more effective.

5.3.1.4 Fully connected layer

The most important layer in a CNN is the fully-connected layer. Every neuron in a fully connected layer is entirely connected with every other neuron in the preceding layer. It is employed at the end of the CNN architecture used for classification. The objective of this layer is to use the features that are identified by the preceding layers and to classify the images. In the current research, the fully connected layer classifies the images into either normal or glaucoma.

CNN algorithms have been extensively used in the field of ophthalmology. To get the intended outcome, a deep learning system needs to be trained. The training process, in general, provides the network with data, analyzes the outcomes, updates the network and continues the process repeatedly until the desired accuracy is attained. Millions of parameters shall be trained since CNNs have several layers. Large ophthalmic image datasets, especially labeled images, are uncommon. Hence transfer learning strategies have been used to overcome this restriction. In transfer learning, a CNN that has been previously trained on a large image dataset is employed for extracting features and classification with a relatively small dataset. Nowadays, CNNs that recognize glaucoma on smaller imaging datasets are trained using the widely employed transfer learning approaches (Weiss *et al.* 2016). In the present research, ORIGA, STARE and REFUGE datasets are used on the pre-trained deep learning algorithms VGG16, Inception V3 and Inception-ResNet-V2 to classify the images into either normal or glaucoma.

5.3.2 VGG-16

VGG-16 is a CNN architecture that is regarded as one of the best vision model architectures. It has convolutional layers, activation functions, pooling layers and fully connected layers. The 16 layers holding weights are represented by the number 16 in VGG. The input goes through a convolutional layer where each filter is applied with a very small receptive field. Max pooling is then followed by the convolutional layers. On a two by two pixel window, max pooling is performed before three completely linked layers. Fully connected layer is followed by the Softmax function which is employed at the output node.

5.3.3 Inception V3

InceptionV3 is another pre-trained neural network architecture employed for image recognition. It typically has two parts. First, the CNN component for feature extraction and second, the fully connected layer with softmax function for classification. Three Inception A modules, five Inception B modules and two Inception C modules are stacked in Inception-v3. There are 48 deep layers in this network. InceptionV3 is the combination of 3 convolutional layers of size 1x1, 3x3 and 5x5 with its outcomes combined into one output vector (Dong *et al.* 2020). A set of convolutional layers employing ReLU as its activation function are then applied. The outcomes of all these layers are applied to the transition layer that concatenates all the convolution layers. The final phase involves the addition of a softmax layer as a classifier that outputs probabilities for each class.

5.3.4 Inception-ResNet-V2

Inception-ResNet-V2 is developed using data from more than a million images in the ImageNet database. It has 164 deep layers. It contains two modules like Inception and Residual (Ferreira *et al.* 2018). As this network includes the benefits of both the residual connections and the inception model, this model is superior (Li *et al.* 2019). The input layer of this network requires an input size of 299x299. In inception, the network employs convolutional filters with various sizes to the same input map, whereas residual modules add input feature maps to a convoluted output. In the current research, this architecture has given the best performance among the other deep learning models.

Pseudocode for the classification of glaucoma using deep learning is shown below.

```
Input: Enhancement image σim
Output: Classified label ψc
Input the Network layers
Input the trained features
Input label
Train Label =70%
Test Label = 30%
Lab=unique (Label)
For ii=1:length (Lab)
        Class =find (Label==Lab(ii))
        // trainfeatures data passed to every layers
          Traincut=length(class)-traincut
          Traindata=[traindata; trainfeatures; class(1:Traincut)]
            Predict Label=classify (net,traindata)
End
For ii=1:size (traindata;1)
      Traindata = [traindata; trainfeatures; class(1:Traincut)]
End
For ii=1:size (trainfeatures;1)
      Traindata = [trainfeatures; trainfeatures; class(1:Traincut)]
End
```

5.4 GLAUCOMA CLASSIFICATION USING ENSEMBLE NETWORK (GCENet)

The aim of the current research is to develop GCENet, a novel ensemble deep learning model that includes three separate deep learning models to improve the classification task for detecting glaucoma. The ensemble method is an efficient technique that includes various independent classification models. Later, it combines all these models to improve accuracy. In the current research, an ensemble of the three distinct deep CNN models like VGG16, Inception V3 and Inception-ResNet-V2 is employed to detect glaucoma. The deep features relevant for the glaucoma diagnosis have been extracted. Finally, the process is completed by directing ensemble characteristics through a classifier layer to differentiate normal images from glaucoma.

In general, integrating different Deep CNNs performs better in classification than a single architecture (Parvin *et al.* 2015). An effective way to maximize the single classification performance would be to build an ensemble of various Deep CNN models. In the present research, an ensemble model, GCENet, is developed for the diagnosis of glaucoma using fundus images. ORIGA, STARE and REFUGE datasets are employed. The three pre-trained deep learning architectures VGG16, Inception V3 and Inception-ResNet-V2 are fine-tuned for the classification of the glaucoma. Deep learning algorithms have given the maximum accuracy than the machine learning algorithms. An ensemble model has been proposed for automatic glaucoma classification to achieve even better results. In the proposed ensemble method, the retinal fundus images are given as the input to the deep learning architectures like VGG16, Inception V3 and Inception-ResNet-V2. Later, these three distinct deep learning architectures are ensembled for the accurate glaucoma classification.

In the current research, Majority Voting (MV) has been employed to ensemble the CNN classifier predictions. MV is used to select the image class based on the decision made by the majority of the models. In this, pi is determined by using Equation (5.4).

$$p_i = \frac{\sum_{j=1}^{m} G(p_{ij})}{m} \quad i = 1,\ldots,n \tag{5.4}$$

Equation (5.4) is normalized using m.

$$G(p_{ij}) = \begin{cases} 1 & \text{if } p_{ij} = \max(CNN_j(x)) \\ 0 & \text{otherwise} \end{cases} \tag{5.5}$$

where p_{ij} represents the output probability, m is the total count of CNN models and n is the output class count. On the three separate data sets, the effectiveness of VGG-16, Inception V3 and Inception-ResNet-V2 and the suggested ensemble architecture is examined. The ensemble model is found to be more accurate and efficient than the individual deep learning models. The ensemble architecture produces the maximum accuracy of 99.6% for REFUGE dataset. The steps involved in the ensemble model are listed below.

Algorithm: GCENet - Ensemble Model

Step1: Divide the images for training and testing processes.

Step 2: Give the images to the three deep learning models VGG16, Inception V3 and Inception-ResNet-V2

Step 3: From the above models, deep features are automatically extracted for glaucoma diagnosis.

Step 4: Integrate the output of the pre-trained neural networks into a prediction vector. Finally, get the results via majority voting.

Step 5: Classify retinal fundus images into either normal or glaucoma.

Initially, the machine learning algorithms, namely Neural Network, SVM and Random Forest are employed. The machine learning algorithms have given the maximum accuracy of 91.8% for ORIGA dataset, 90.6% for STARE dataset and 93.4% for REFUGE dataset. Secondly, the three pre-trained deep learning architectures VGG16, Inception V3 and Inception-ResNet-V2 are fine-tuned for the classification of the glaucoma. The deep learning algorithms have given the maximum accuracy of 98.9% for ORIGA, 98.3% for STARE and 99.1% for REFUGE datasets. Thirdly, an ensemble model has been proposed for automatic glaucoma classification to achieve even better results. It is observed that the accuracy is improved significantly by the majority voting based ensemble method. Moreover, the suggested ensemble classifier for automatic glaucoma detection is found to be more accurate and robust than the individual optimized deep CNN models.

5.5 EVALUATION METRICS FOR GLAUCOMA CLASSIFICATION

In the current research, the efficacy of the recommended method is calculated by applying the assessment metrics like accuracy, sensitivity and specificity. The evaluation metrics are represented below.

5.5.1 Accuracy

Accuracy depicts the percentage of right predictions produced.

$$Accuracy = \frac{TP + TN}{TP + FP + TN + FN} \quad (5.6)$$

5.5.2 Sensitivity

The real positive rate is represented as sensitivity and it evaluates the successful prediction of the positive class.

$$Sensitivity = \frac{TP}{TP + FN} \tag{5.7}$$

5.5.3 Specificity

The real negative rate is represented as specificity and it evaluates the successful prediction of the negative class.

$$Specificity = \frac{TN}{TN + FP} \tag{5.8}$$

True Positive (TP) in the equations above is the number of images that are correctly identified as having glaucoma and the True Negative (TN) is the number of images that are correctly identified as having normal vision. False Negative (FN) represents the number of glaucoma images categorized as normal and False Positive (FP) refers to the number of normal images recognized as glaucoma images.

5.6 SUMMARY

A widely available, economically screening strategy, is needed due to the rising prevalence of glaucoma, and the proposed model may be helpful for the early diagnosis of glaucoma for the people who are affected by this disease. The current research focuses on the investigation of both the machine learning and the deep learning algorithms for the detection of glaucoma. Research has been carried out to find the best suited method for the diagnosis of glaucoma. The GCENet, an ensemble model is developed on the three separate deep CNNs for the diagnosis of glaucoma. Automated diagnosis of glaucoma constitutes 5 phases such as image preprocessing, segmentation, feature extraction, image classification and performance evaluation as described. The deep learning algorithms could significantly enhance the performance of glaucoma diagnosis.

CHAPTER 6

RESULTS AND DISCUSSION

The outcomes of the segmentation, feature extraction and classification techniques utilizing the machine learning and deep learning algorithms are presented in this chapter.

6.1 RESULTS OF SEGMENTATION

The performance measures used for the evaluation of segmentation in the current research are IoU and Dice score. These metrics give a prompt assessment of segmentation from the U-Net model. Table 6.1 shows the results of the optic disc segmentation for ORIGA, STARE and REFUGE datasets whereas Table 6.2 shows the results of the optic cup segmentation for ORIGA, STARE and REFUGE datasets.

Table 6.1 Results of Optic Disc Segmentation

U-Net Metrics	ORIGA	STARE	REFUGE
Dice Score	0.89	0.87	0.94
IoU	0.85	0.84	0.93

Table 6.2 Results of Optic Cup Segmentation

U-Net Metrics	ORIGA	STARE	REFUGE
Dice Score	0.86	0.84	0.91
IoU	0.83	0.81	0.90

The model is assessed on ORIGA, STARE and REFUGE datasets and a maximum Dice score of 0.89, 0.87 & 0.94 and maximum IoU of 0.85, 0.84 & 0.93 are obtained. It can be observed from the results that the optic disc and the optic cup segmentations have the best dice score and IoU for REFUGE dataset. The results confirm that the segmentation approach is capable of working efficiently with unknown images.

6.2 RESULTS OF GLAUCOMA CLASSIFICATION USING MACHINE LEARNING

Classification of glaucoma has been evaluated for ORIGA, STARE and REFUGE datasets for both with and without PCA. The classification results for the same have been presented in Tables 6.3, 6.4 and 6.5.

Table 6.3 Results of Glaucoma Classification for ORIGA Dataset

Classifier	Before/After PCA	Sensitivity (%)	Specificity (%)	Accuracy (%)
Neural Network	Before PCA	81.6	80.9	83.7
	After PCA	83.7	81.4	86.5
SVM	Before PCA	84.8	85.4	87.5
	After PCA	91.2	89.6	90.3
Random Forest	Before PCA	91.6	87.8	88.4
	After PCA	92.7	90.1	91.8

Table 6.4 Results of Glaucoma Classification for STARE Dataset

Classifier	Before/After PCA	Sensitivity (%)	Specificity (%)	Accuracy (%)
Neural Network	Before PCA	76.7	78.4	81.3
	After PCA	78.9	80.7	84.6
SVM	Before PCA	88.3	84.8	85.2
	After PCA	90.4	86.5	89.4
Random Forest	Before PCA	88.4	85.6	87.7
	After PCA	89.7	87.3	90.6

Table 6.5 Results of Glaucoma Classification for REFUGE Dataset

Classifier	Before/After PCA	Sensitivity (%)	Specificity (%)	Accuracy (%)
Neural Network	Before PCA	82.9	81.2	85.8
	After PCA	84.6	82.7	88.3
SVM	Before PCA	90.5	87.4	89.4
	After PCA	93.1	89.8	91.6
Random Forest	Before PCA	89.7	88.4	90.2
	After PCA	92.8	90.4	93.4

Experimental results show that, the selection of optimal features significantly change the classification accuracy. The complexity is reduced and unnecessary features are eliminated by using PCA. Thus, by reducing the irrelevant features, the accuracy of glaucoma diagnosis has been increased. It is observed that, for the REFUGE dataset, Random forest acquires the maximum accuracy of 93.4% after the process of PCA. It is perceived that the accuracy of glaucoma classification has been improved when PCA is used.

For ORIGA dataset, the proposed model attains a maximum classification accuracy of 91.8%, sensitivity of 92.7% and specificity of 90.1%. For STARE dataset, the proposed model attains a maximum classification accuracy of 90.6%, sensitivity of 89.7% and specificity of 87.3%. In REFUGE dataset, a maximum accuracy of 93.4%, sensitivity of 92.8% and specificity of 90.4% have been achieved. The outcomes from the datasets are also plotted in the form of graphs. Figure 6.1 shows the performance of the machine learning algorithms such as Neural Network, SVM and Random Forest for ORIGA, STARE and REFUGE datasets.

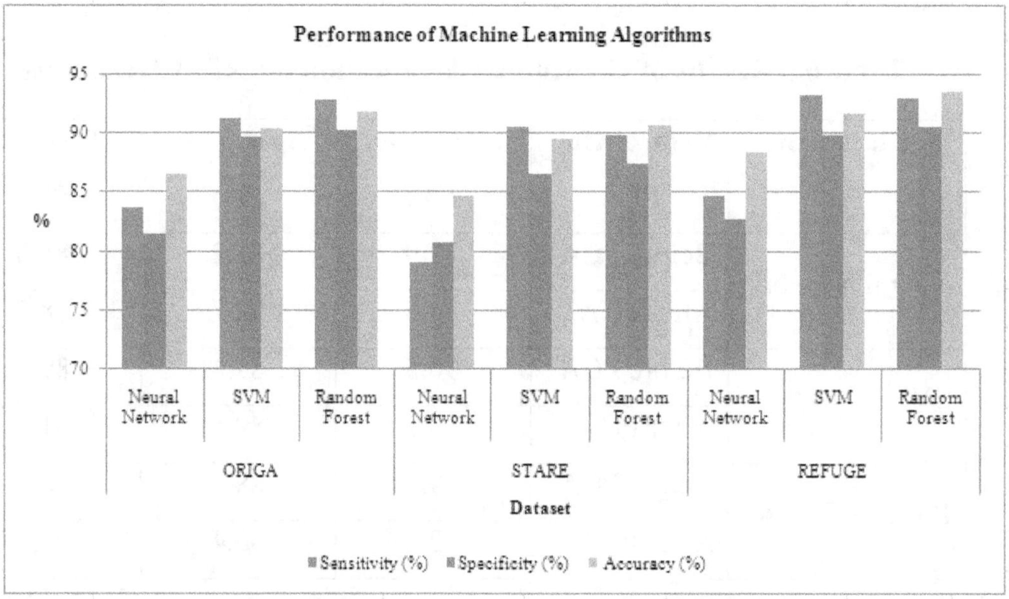

Figure 6.1 Performance of Machine Learning algorithms

The results obtained from the machine learning algorithms can still be improved to obtain an accurate glaucoma classification model. Deep learning is then used to improve the performance of the model.

6.3 RESULTS OF GLAUCOMA CLASSIFICATION USING DEEP LEARNING

For ORIGA dataset, the proposed model obtains a maximum classification accuracy of 98.9%, sensitivity of 99.6% and specificity of 98.9%, whereas for STARE dataset, it attains a maximum classification accuracy of 98.3%, sensitivity of 99.3% and specificity of 97.5%. Likewise, for REFUGE dataset, a maximum accuracy of 99.1%, sensitivity of 99.7% and specificity of 98.2% have been achieved.

Table 6.6 and Figure 6.2 show the performance of the deep learning algorithms such as VGG16, Inception V3 and Inception-ResNet-V2 for ORIGA, STARE and REFUGE datasets.

Table 6.6 Performance of Deep Learning Algorithms

Dataset	Deep Learning Algorithms	Sensitivity (%)	Specificity (%)	Accuracy (%)
ORIGA	VGG16	94.4	93.4	94.5
	Inception V3	96.8	94.1	96.3
	Inception-ResNet-V2	99.6	98.9	98.9
STARE	VGG16	93.3	89.2	93.4
	Inception V3	94.2	93.4	95.6
	Inception-ResNet-V2	99.3	97.5	98.3
REFUGE	VGG16	95.3	93.8	94.7
	Inception V3	96.9	94.7	97.4
	Inception-ResNet-V2	99.7	98.2	99.1

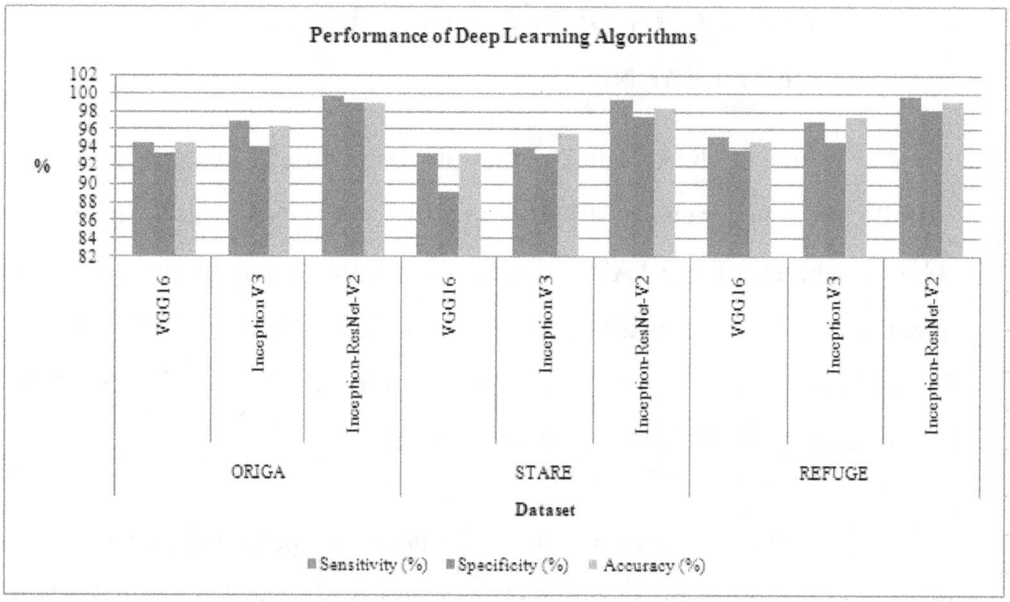

Figure 6.2 Performance of Deep Learning algorithms

From Figures 6.1 and 6.2, it can be observed that the machine learning algorithms have given the maximum accuracy of 91.8% for ORIGA dataset, 90.6% for STARE dataset and 93.4% for REFUGE dataset. On the contrary, deep learning algorithms have given the maximum accuracy of 98.9% for ORIGA, 98.3% for STARE and 99.1% for REFUGE datasets. From this analysis, it is seen that accuracy has been improved by using deep learning algorithms. From the results, it can be concluded that Random Forest algorithm performs better among the different machine learning algorithms and Inception-ResNet-V2 is more effective and accurate in detecting the glaucoma than the other deep learning algorithms.

6.4 RESULTS OF GLAUCOMA CLASSIFICATION USING ENSEMBLE DEEP LEARNING MODEL

It is observed from the results that the GCENet ensemble classifier for automatic detection of glaucoma is more accurate and robust than the individual optimized deep CNN models. For ORIGA dataset, the proposed ensemble model obtains a maximum classification accuracy of 99.3%, sensitivity of 99.8% and specificity of 99.4%, whereas for STARE dataset, it attains a maximum classification accuracy of 99.5%, sensitivity of 99.6% and specificity of 98.4%. However, for REFUGE dataset, a maximum accuracy of 99.6%, sensitivity of 99.9% and specificity of 99.5% have been achieved. Table 6.7 shows the performance of the ensemble model for ORIGA, STARE and REFUGE datasets.

Table 6.7 Performance of Ensemble Model

Dataset	Sensitivity (%)	Specificity (%)	Accuracy (%)
ORIGA	99.8	99.4	99.3
STARE	99.6	98.4	99.5
REFUGE	99.9	99.5	99.6

It can be observed from the results that the combination of the classifiers in the ensemble model has produced a good glaucoma classification. GCENet is very significant and it has the potential to be used in the real-world detection of glaucoma.

6.5 ANALYSIS OF GLAUCOMA CLASSIFICATION RESULTS

The effectiveness of VGG16, Inception V3, Inception-ResNet-V2 and the suggested ensemble architecture are examined on three different datasets. Table 6.8 shows the performance of deep learning and ensemble model for ORIGA, STARE and REFUGE datasets. It can be noticed from Table 6.8 and Figures 6.3 that the proposed ensemble model has achieved good performance than the traditional methods for the ORIGA, STARE and REFUGE datasets.

Table 6.8 Results of Glaucoma Classification using Deep Learning and Ensemble Model

Dataset	Deep Learning Algorithms	Sensitivity (%)	Specificity (%)	Accuracy (%)
ORIGA	VGG16	94.4	93.4	94.5
	Inception V3	96.8	94.1	96.3
	Inception-ResNet-V2	99.6	98.9	98.9
	Proposed Algorithm	99.8	99.4	99.3
STARE	VGG16	93.3	89.2	93.4
	Inception V3	94.2	93.4	95.6
	Inception-ResNet-V2	99.3	97.5	98.3
	Proposed Algorithm	99.6	98.4	99.5
REFUGE	VGG16	95.3	93.8	94.7
	Inception V3	96.9	94.7	97.4
	Inception-ResNet-V2	99.7	98.2	99.1
	Proposed Algorithm	99.9	99.5	99.6

The comparison of the proposed method with the existing methods is shown in Table 6.9. The existing systems used several transfer learning models and datasets. The comparison appears to be more difficult due to the diversity in the combinations of dataset and transfer learning techniques. The proposed model achieves the maximum accuracy of 99.6%. The outcome of the research shows that the proposed method performs better than the other conventional techniques.

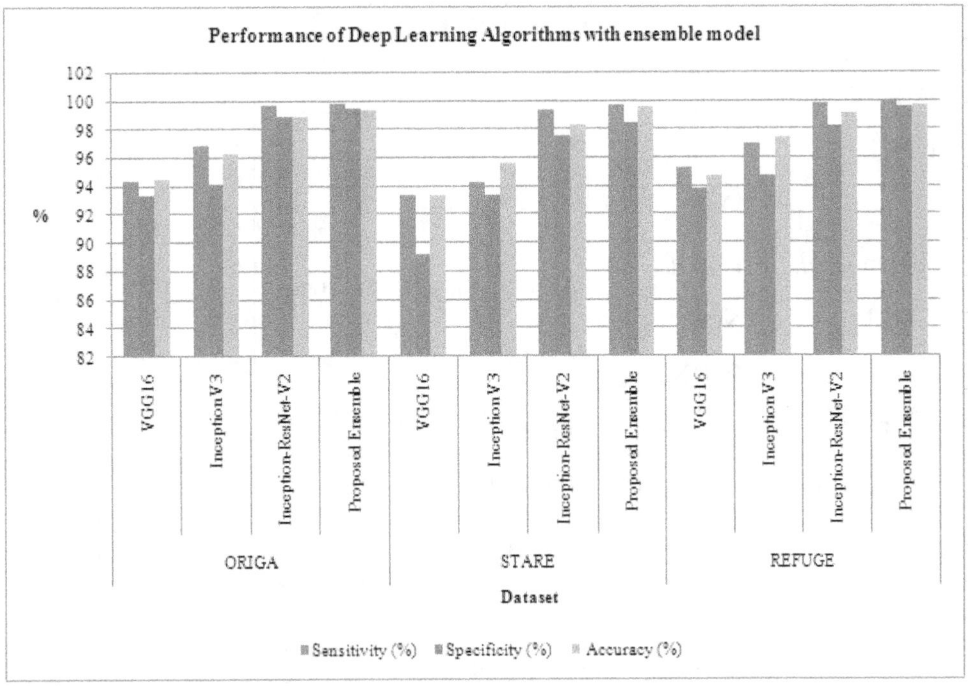

Figure 6.3 Performance of Deep Learning and Ensemble model

Table 6.9 Comparison of the proposed and the existing methods

Author & Year	Methods	Dataset	Accuracy	Sensitivity	Specificity
Raghavendra *et al.* (2018)	18 Layers CNN	Private	98.13%,	98%	98.30%
Li *et al.* (2020)	AG-CNN	LAG RIM-ONE	96.2%	95.4%	96.7%
Chai *et al.* (2018)	MB-NN	Private	91.51%	92.33%	90.90%
Singh *et al.* (2019)	Inception-v3	RIM –ONE	92.51%	92.31%	93.51%
Christopher *et al.* (2018)	VGG16, ResNet50 and Inception v3	Private	91%	88%	95%.

Table 6.9 (Continued)

Ahn *et al.* (2018)	CNN	Private	92.2 %	-	-
Ajesh *et al.* (2021)	Deep Belief Network (DBN)	MESSIDOR	95.9%	88.3%	95.12%
GCENet (Proposed Model)	Ensemble	ORIGA STARE REFUGE	99.6%	99.9%	99.5%

Figure 6.4 shows the accuracy of the proposed ensemble model and the other machine learning and deep learning algorithms. It is observed from the results that the ensemble architecture gives better result than the other models.

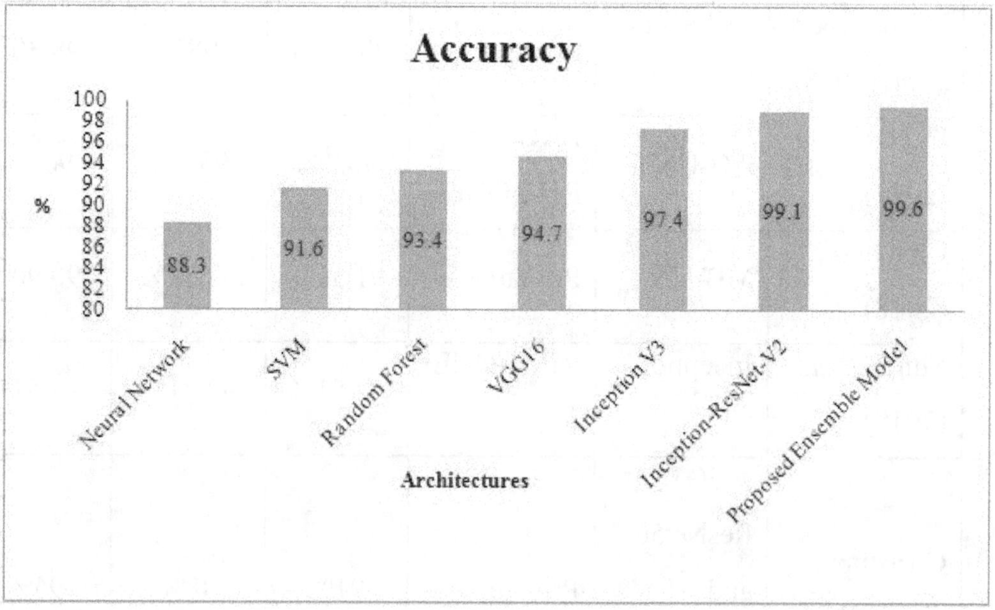

Figure 6.4 Accuracy of the proposed Ensemble model and the other Machine Learning and Deep Learning algorithms

6.6 SUMMARY

In this chapter, the results obtained from the segmentation and classification process have been described. Optic disc and optic cup segmentation have the best dice score and IoU for REFUGE dataset. The selection of optimal features significantly changes the classification accuracy. By reducing the irrelevant features, the accuracy of glaucoma diagnosis has been increased. It is observed that, for the REFUGE dataset, Random forest acquires the maximum accuracy of 93.4% after the process of PCA. On the contrary, deep learning algorithms have given the maximum accuracy of 99.1% for REFUGE dataset. It is observed that the accuracy has been improved by using ensemble model. Thus the combination of the classifiers in the ensemble model has produced a good performance for the classification of glaucoma.

CHAPTER 7

CONCLUSION AND FUTURE ENHANCEMENT

7.1 CONCLUSION

Doctors and surgeons detect glaucoma more precisely due to the advent of computer aided diagnosis. There are several methods for detecting glaucoma, but many have not given high accuracy. As a result, a unique technique for the early detection of glaucoma is necessary. The present research has aimed at developing a glaucoma classification model. Medical diagnostics for glaucoma detection can be made more accurate and efficient with the proposed method. The proposed model helps in the early detection of glaucoma.

The publicly available ORIGA, STARE and REFUGE datasets are used. The images have been preprocessed through Gaussian filter and Histogram Equalization. The preprocessed images are given as the input to the segmentation process. Segmentation has been done based on a semantic segmentation approach using U-Net architecture. The optic disc segmentation has best dice score of 0.94 and IoU of 0.93 for REFUGE dataset. The features are extracted using correlation based method. Additionally, the complexity has been reduced and unnecessary features are eliminated by using PCA. Thus, by reducing the irrelevant features, the accuracy of glaucoma diagnosis has been increased. It is observed that, for the REFUGE dataset, Random forest acquires the maximum accuracy of 90.2% and 93.4% before and after the process of PCA respectively.

The classification has been done through the use of machine learning algorithms such as Neural Network, SVM and Random Forest. Machine learning algorithms have given the maximum accuracy of 91.8% for ORIGA, 90.6% for STARE and 93.4% for REFUGE datasets. The results obtained from the machine learning algorithms are improved by using deep learning algorithms to obtain accurate glaucoma classification model.

Additionally, the retinal fundus images are given as the input to the deep learning architectures such as VGG16, Inception V3 and Inception-ResNet-V2. The deep learning method does not involve manual feature extraction process, which, in turn, makes additional research easy and minimizes the effort of feature analysis. The deep learning algorithms have given the maximum accuracy of 98.9% for ORIGA dataset, 98.3% for STARE dataset and 99.1% for REFUGE dataset.

The GCENet, an ensemble model also has been proposed for automatic glaucoma classification in order to achieve even better results. In the proposed ensemble method, the retinal fundus images are given as the input to the deep learning architectures, namely VGG16, Inception V3 and Inception-ResNet-V2. Later, these three distinct deep learning architectures are ensembled for the accurate classification of glaucoma. For ORIGA dataset, the proposed ensemble model obtains a maximum classification accuracy of 99.3%, sensitivity of 99.8% and specificity of 99.4%, whereas for STARE dataset, it attains a maximum classification accuracy of 99.5%, sensitivity of 99.6% and specificity of 98.4%. For REFUGE dataset, a maximum accuracy of 99.6%, sensitivity of 99.9% and specificity of 99.5% have been achieved. It is observed that accuracy has been significantly improved by using the majority voting based ensemble method. It is observed from the results that the suggested ensemble classifier for automatic glaucoma detection is more accurate and robust than the individual optimized deep

CNN models. The proposed ensemble algorithm outperforms the other existing methods. The ensemble model of multiple classifiers provides an added advantage in producing great outcomes. The proposed model shows significant performance and hence it can be used as a suitable tool for glaucoma diagnosis. It has the potential to be used in the real-world for the detection of glaucoma. It can help medical professionals and clinicians in assessing the risk of glaucoma in patients.

7.2 FUTURE ENHANCEMENT

In the future, the research related to glaucoma should concentrate on tilted discs which can be done through existing or new texture descriptors that help in automatically adapting the variability found in the textures. The fundus image analysis model with an embedded processor can be inserted on the fundus image-taking camera. In this process, taking images from that camera itself indicates the seriousness of the disease in the human being. The processor embedded in the fundus camera does not require any trained person to monitor the screening zone. The accuracy could be further increased in the future by using a larger training sample size, additional feature choices and feature combinations. Even this proposed approach can be extended to the other eye-related problems like age-related macular degeneration, macular edema and hypertensive retinopathy diseases. The model can also be installed on a mobile device as a Mobile App and utilized by clinicians for diagnosis. This deployment could be a part of future work.

REFERENCES

1. Abdel-Hamid, L 2020, 'Glaucoma detection from retinal images using statistical and textural wavelet features'. Journal of Digital Imaging, vol. 33, no. 1, pp. 151–158.

2. Acharya, U Rajendra & Dua, Sumeet & du, Xian & Subbhuraam, Vinitha Sree & Chua, Kuang 2011, 'Automated Diagnosis of Glaucoma Using Texture and Higher Order Spectra Features. Information Technology in Biomedicine', IEEE Transactions, vol. 15, pp. 449 - 455. 10.1109/TITB.2011.2119322.

3. Acharya, UR., Bhat, S, Koh, JE W, Bhandary, SV & Adeli, H 2017, 'A novel algorithm to detect glaucoma risk using texton and local configuration pattern features extracted from fundus images'. Computers in Biology and Medicine, 88, 72–83.

4. Afolabi, OJ, Mabuza-Hocquet, GP, Nelwamondo, FV & Paul, BS 2021, 'The use of U-Net lite and Extreme Gradient Boost (XGB) for glaucoma detection'. IEEE Access, vol. 9, pp. 47411-47424.

5. Ahmad, H, Yamin, A, Shakeel, A, Gillani, SO & Ansari, U 2014, 'Detection of glaucoma using retinal fundus images'. In 2014 International conference on robotics and emerging allied technologies in engineering (iCREATE), pp. 321-324. IEEE.

6. Ahn, JM, Kim, S, Ahn, KS, Cho, SH, Lee, KB & Kim, US 2018, 'A deep learning model for the detection of both advanced and early glaucoma using fundus photography', PloS one, vol. 13, no. 11, pp. e0207982.

7. Ajesh, F, Ravi, R & Rajakumar, G 2021, 'Early diagnosis of glaucoma using multi-feature analysis and DBN based classification', Journal of Ambient Intelligence and Humanized Computing, vol. 12, no. 3, pp. 4027-4036.

8. Akbar S, Hassan SA, Shoukat A, Alyami J, Bahaj SA. Detection of microscopic glaucoma through fundus images using deep transfer learning approach. Microsc Res Tech. 2022 Jun;85(6):2259-2276. doi: 10.1002/jemt.24083. Epub 2022 Feb 15. PMID: 35170136.

9. Al-Bander, Baidaa, Williams, Bryan, Al-Nuaimy, Waleed, Al-Taee, Majid, Pratt, Harry, Zheng & Yalin 2018, 'Dense Fully Convolutional Segmentation of the Optic Disc and Cup in Colour Fundus for Glaucoma Diagnosis'. Symmetry. 10. 87. 10.3390/sym10040087.

10. Ali, Riaz, Li, Ping, Chen, Yan, Li, Huating, Yang, Po, Jung, Younhyun, Kim, Jinman & Chen, C 2020, 'Optic Disc and Cup Segmentation Through Fuzzy Broad Learning System for Glaucoma Screening'. IEEE Transactions on Industrial Informatics, DOI: 10.1109/TII.2020.3000204.

11. Allison, K, Patel, D & Alabi, O 2020, 'Epidemiology of Glaucoma: The Past, Present, and Predictions for the Future. Cureus', vol. 12, no. 11, pp. e11686. https://doi.org/10.7759/cureus.11686

12. Almazroa, A, Burman, R, Raahemifar, K & Lakshminarayanan, V 2015, 'Optic Disc and Optic Cup Segmentation Methodologies for Glaucoma Image Detection: A Survey', Journal of Ophthalmology, vol. 2015.

13. Al-Naser, M & Soderstrom 2012, 'Reconstruction of occluded facial images usingasymmetrical principal component analysis', Integr. Comput.-Aided Eng. vol. 19, no. 3, pp. 273–283.

14. Aquino, A, Gegundez-Arias, ME & Marin, D 2010, 'Detecting the optic disc boundary in digital fundus images using morphological, edge detection, and feature extraction techniques'. IEEE Trans Med Imaging. vol. 29, no. 11, pp. 1860-9. doi: 10.1109/TMI.2010. 2053042. Epub 2010 Jun 17. PMID: 20562037

15. Armstrong, R & Cubbidge, R 2014, The eye and vision: an overview. In V. Preedy (Ed.), Handbook of nutrition, diet, and the eye, Academic Press. https://doi.org/10.1016/B978-0-12-401717-7.00001-0, pp. 3-10.

16. Asaoka, R, Hirasawa, K, Iwase, A, Fujino, Y, Murata, H, Shoji, N & Araie, M 2017, 'Validating the usefulness of the "random forests" classifier to diagnose early glaucoma with optical coherence tomography'. American Journal of Ophthalmology, vol. 174, pp. 95–103.

17. Bach Cuadra, M, Duay, V & Thiran, JP 2015, 'Atlas-based Segmentation. In: Paragios, N, Duncan, J, Ayache, N (eds) Handbook of Biomedical Imaging. Springer, Boston, MA. https://doi.org/10.1007/978-0-387-09749-7_12

18. Bankhead, CN, Scholfield, JG, McGeown & Curtis, TM 2012, 'Fast retinal vessel detection and measurement using wavelets and edge location refinement', PLoS ONE, vol. 7, no. 3, Art. Number: e32435

19. Chen, HSL, Chen, GA, Syu, JY, Chuang, LH, Su, WW, Wu, WC & Kang, EYC 2022, 'Early Glaucoma Detection By Using Style-Transfer to Predict Retinal Nerve Fiber Layer Thickness Distribution on Fundus Photograph'. Ophthalmology Science, pp. 100180.

20. Cheng J 2013, 'Superpixel Classification Based Optic Cup Segmentation. In: Mori K, Sakuma I, Sato, Y, Barillot, C, Navab, N (eds) Medical Image Computing and Computer-Assisted Intervention – MICCAI 2013. MICCAI 2013. Lecture Notes in Computer Science, vol. 8151. Springer, Berlin, Heidelberg. https://doi.org/10.1007/978-3-642-40760-4_53

21. Chiranji Lal Chowdhary, DP & Acharjya 2020, 'Segmentation and Feature Extraction in Medical Imaging: A Systematic Review', Procedia Computer Science, vol. 167, pp. 26-36, ISSN 1877-0509, https://doi.org/10.1016/j.procs.2020.03.179.

22. Christopher, M, Belghith, A, Bowd, C, Proudfoot, JA, Goldbaum, MH, Weinreb, RN & Zangwill, LM 2018, 'Performance of deep learning architectures and transfer learning for detecting glaucomatous optic neuropathy in fundus photographs', Scientific reports, vol. 8, no. 1, pp. 1-13.

23. Cootes, CJ, Taylor, DH, Cooper, J & Graham 1995, 'Active Shape Models-Their Training and Application', Computer Vision and Image Understanding, vol. 61, issue. 1, pp. 38-59, ISSN 1077-3142, https://doi.org/10.1006/cviu.1995.1004.

24. Cortes, C & Vapnik, V 1995, 'Support-Vector Networks. Machine Learning, vol. 20, pp. 273-297. https://doi.org/10.1007/BF00994018

25. Damon, W. W. K., Liu, J., Meng, T. N., Fengshou, Y., & Yin, W. T. (2012, May). 'Automatic detection of the optic cup using vessel kinking in digital retinal fundus images. In 2012 9th IEEE international symposium on biomedical imaging (ISBI) (pp. 1647-1650). IEEE.

26. Davies, ER 2004, Machine Vision, Third edition, Morgan Kaufmann publications.

27. Devecioglu, OC, Malik, J, Ince, T, Kiranyaz, S, Atalay, E & Gabbouj, M 2021, 'Real-time glaucoma detection from digital fundus images using self-onns'. IEEE Access, 9, 140031-140041.

28. Diaz-Pinto A, Morales S, Naranjo V, Köhler T, Mossi JM & Navea A 2019, 'CNNs for automatic glaucoma assessment using fundus images: An extensive validation', Biomed. Eng. OnLine, vol. 18, no. 1, pp. 29.

29. Dong, N, Zhao, L, Wu, CH & Chang, JF 2020, 'Inception v3 based cervical cell classification combined with artificially extracted features'. Appl Soft Comput , vol. 93, pp. 106311

30. Dua, Sumeet, Acharya, U Rajendra, Chowriappa, Pradeep, Subbhuraam, & Vinitha Sree 2011, 'Wavelet-Based Energy Features for Glaucomatous Image Classification'. IEEE Transactions on Information Technology in Biomedicine. vol. 16. pp. 80-87. 10.1109/TITB.2011.2176540.

31. Fakhira Zahra Zulfira, Suyanto Suyanto & Anindita Septiarini 2021, 'Segmentation technique and dynamic ensemble selection to enhance glaucoma severity detection', Computers in Biology and Medicine, vol. 139, pp. 104951, ISSN 0010-4825, https://doi.org/10.1016/j.compbiomed.2021.104951.

32. Ferreira, CA, Melo, T, Sousa, P, Meyer, MI, Shakibapour, E, Costa, P, Campilho 2018, 'Classification of Breast Cancer Histology Images Through Transfer Learning Using a Pre-trained Inception Resnet V2'. In Proceedings of the International Conference Image Analysis and Recognition, Póvoa de Varzim, Portugal, Chapter 86. pp. 763–770. [CrossRef]

33. Fu, H, Cheng, J, Xu, Y, Wong, DWK, Liu, J & Cao, X 2018, 'Joint optic disc and cup segmentation based on multi-label deep network and polar transformation'. IEEE Transactions on Medical Imaging, vol. 37, no. 7, pp. 1597–1605.

34. Fu, Huazhu, Xu, Yanwu, Lin, Stephen, Wong, Damon, Mani, Baskaran, Mahesh, Meenakshi, Aung, Tin & Liu, Jiang 2018, 'Multi-context Deep Network for Angle-Closure Glaucoma Screening in Anterior Segment OCT'. 10.1007/978-3-030-00934-2_40.

35. Fumero, F, Alayón, S, Sanchez, JL, Sigut, J & Gonzalez-Hernandez, M 2011, ''RIM-ONE: An open retinal image database for optic nerve evaluation', in Proc. 24th Int. Symp. Comput.-Based Med. Syst. (CBMS), pp. 1–6.

36. Gomez-Valverde, Juan, Antón, Alfonso, Fatti, Gianluca, Liefers, Bart, Herranz, Alejandra, Santos, Andres, Sánchez, Clara, Ledesma-Carbayo, & María 2019, 'Automatic glaucoma classification using color fundus images based on convolutional neural networks and transfer learning'. Biomedical Optics Express. 10. 892. 10.1364/BOE.10.000892.

37. Gopika, N & Kowshalaya, AM 2018, 'Correlation Based Feature Selection Algorithm for Machine Learning', 3rd International Conference on Communication and Electronics Systems (ICCES), pp. 692-695, doi: 10.1109/CESYS.2018.8723980.

38. Guo, F, Li, W & Tang, J 2020, 'Automated glaucoma screening method based on image segmentation and feature extraction'. Med Biol Eng Comput, vol. 58, pp. 2567–2586, https://doi.org/10.1007 /s11517-020-02237-2

39. Hancox, MD 1999, 'Optic disc size, an important consideration in the glaucoma evaluation', Clinical Eye and Vision Care, vol. 11, no. 2, pp. 59 – 62.

40. Hnoohom, N & Jitpattanakul, A 2017, 'Comparison of Ensemble Learning Algorithms for Cataract Detection from Fundus Images'. 21st International Computer Science and Engineering Conference (ICSEC), pp. 1-5.

41. Hoover A & Goldbaum, M 2003, 'Locating the optic nerve in a retinal image using the fuzzy convergence of the blood vessels', IEEE Transactions Med. Imag., vol. 22, no. 8, pp. 951–958.

42. Imani, E, Javidi, M & Pourreza, HR 2015, 'Improvement of retinal blood vessel detection using morphological component analysis', Computer methods and programs in biomedicine, vol. 118, no. 3, pp. 263-279.

43. Islam, MT, Mashfu, ST, Faisal, A, Siam, SC, Naheen, IT & Khan, R 2021, 'Deep learning-based glaucoma detection with cropped optic cup and disc and blood vessel segmentation'. IEEE Access, vol. 10, pp. 2828-2841.

44. Jonas JB, Aung T, Bourne RR, Bron AM, Ritch R, Panda-Jonas S. Glaucoma. Lancet. 2017 Nov 11;390(10108):2183-2193. doi: 10.1016/S0140-6736(17)31469-1. Epub 2017 May 31. PMID: 28577860.

45. Jonas, JB, Bergua, A, Schmitz Valckenberg, P, Papastathopoulos, KL & Budde, WM 2000, 'Ranking of optic disc variables for detection of glaucomatous optic nerve damage', Invest. Ophthalmol. Vis. Sci., vol. 41, no. 7, pp. 1764–1773.

46. Jonas, JB, Mardin, CY & Grundler, AE 1998, 'Comparison of measurements of neuroretinal rim area between confocal laser scanning tomography and planimetry of photographs', Br. J. Ophthalmol., vol. 82, no. 4, pp. 362–366.

47. Joshi, JD, Sivaswamy, J & Krishnadas, JL 2011, 'Optic Disk and Cup Segmentation from Monocular Colour Retinal Images for Glaucoma Assessment', IEEE Trans. Med. Imag., vol. 30, no. 6, pp. 1192–1205.

48. Julesz, B 1981, 'Textons the elements of texture perception, and their interactions'. Nature, vol. 290, pp. 91–97, https://doi.org/ 10.1038/ 290091 a0 arXiv:1011.1669v3.

49. Kaur, J & Mittal, D 2017, 'A generalized method for the detection of vascular structure in pathological retinal images', Biocybernetics and Biomedical Engineering, vol. 37, no. 1, pp. 184-200.

50. Kaur, P & Khosla, PK 2020, 'Artificial Intelligence Based Glaucoma Detection' In: Verma O., Roy S., Pandey S., Mittal M. (eds) Advancement of Machine Intelligence in Interactive Medical Image Analysis. Algorithms for Intelligent Systems. Springer, Singapore., https://doi.org/10.1007/978-981-15-1100-4_14

51. Kausu, T, Gopi, VP, Wahid, KA, Doma, W & Niwas, SI 2018, 'Combination of clinical and multiresolution features for glaucoma detection and its classification using fundus images', Biocybern. Biomed. Eng. vol. 38 no. 2, pp. 329–341.

52. Kim, PY, Iftekharuddin, KM, Davey, PG, Tóth, M, Garas, A, Holló, G & Essock, EA 2013, 'Novel fractal feature-based multiclass glaucoma detection and progression prediction'. IEEE J Biomed Health Inform. vol. 17, no. 2, pp. 269-276. doi: 10.1109/TITB.2012 2218661. Epub 2012 Sep 12. PMID: 22987538.

53. Kingman & Sharon 2004, 'Glaucoma is second leading cause of blindness globally'. Bulletin of the World Health Organization. vol. 82, pp. 887-888. 10.1590/S0042-96862004001100019.

54. Ko, M, Leung, C & Yuen, T 2016, 'Automated segmentation of optic nerve head for the topological assessment', in Proc. Global Med. Eng. Phys. Exchanges/Pan Amer. Health Care Exchanges (GMEPE/PAHCE), pp. 1–4

55. Kumar, G & Bhatia, PK 2014, 'A Detailed Review of Feature Extraction in Image Processing Systems', Fourth International Conference on Advanced Computing & Communication Technologies, pp. 5-12, doi: 10.1109/ACCT.2014.74

56. Lee, K, Niemeijer, M, Garvin, MK, Kwon, YH, Sonka, M & Abramoff, MD 2010, 'Segmentation of the optic disc in 3-D OCT scans of the optic nerve head', IEEE Trans. Med. Imag., vol. 29, no. 1, pp. 159–168.

57. Lei, H, Liu, W, Xie, H, Zhao, B, Yue, G & Lei, B 2021, 'Unsupervised domain adaptation based image synthesis and feature alignment for joint optic disc and cup segmentation'. IEEE Journal of Biomedical and Health Informatics, vol. 26, no. 1, pp. 90-102.

58. Li, H, Hu, M & Huang, Y 2019, 'Automatic identification of overpass structures: a method of deep learning'. ISPRS Int J Geo Inf, vol. 8, no. 9, pp. 421.

59. Li, L, Xu, M, Liu, H, Li, Y, Wang, X, Jiang, L, Wang, Z, Fan, X & Wang, N 2020, 'A Large-Scale Database and a CNN Model for Attention-Based Glaucoma Detection'. IEEE Trans Med Imaging. vol. 39, no. 2, pp. 413-424. doi: 10.1109/TMI.2019.2927226. Epub 2019 Jul 8. PMID: 31283476.

60. Li, Z, He, Y, Keel, S, Meng, W, Chang, RT & He, M 2018, 'Efficacy of a deep learning system for detecting glaucomatous optic neuropathy based on color fundus photographs'. Ophthalmology, vol. 125, no. 8, pp. 1199–1206.

61. Lim, WS, Chong, MS & Sahadevan, S 2007, Utility of the clinical dementia rating in Asian populations. Clinical medicine & research, vol. 5, no. 1, pp. 61–70. https://doi.org/10.3121/cmr.2007.693

62. Mohamed, SM, Wahba, AA & Eldosoky, MAA 2020, 'Eye Diseases Monitoring Using Statistical Analysis of Disc Topography and RNFL Circular Tomogram Parameters'. In Journal of Physics: Conference Series, IOP Publishing, vol. 1447, no. 1, p. 012010.

63. Mahmood, MT & Lee, IH 2022, 'Optic disc localization in fundus images through accumulated directional and radial blur analysis'. Computerized Medical Imaging and Graphics, vol. 98, pp. 102058.

64. Malik, FM, Batool, F, Rubab, A, Chaudhary, NA, Khan, KB & Qureshi, MA 2020, 'Retinal disorder as a biomarker for detection of human diseases', 2020 IEEE 23rd International Multitopic Conference (INMIC), pp. 1-6, doi: 10.1109/INMIC50486.2020.931 8059.

65. Nayak, DR, Das, D, Majhi, B, Bhandary, SV & Acharya, UR 2021, 'ECNet: An evolutionary convolutional network for automated glaucoma detection using fundus images'. Biomedical Signal Processing and Control, vol. 67, pp. 102559.

66. Neale, RE, Purdie, JL, Hirst, LW & Green, AC 2003, 'Sun exposure as a risk factor for nuclear cataract'. Epidemiology. vol. 14, no. 6, pp. 707–12. doi:10.1097/01.ede.0000086881.84657.98. PMID 14569187. S2CID 40041207.

67. Odstrcilik, J, Kolar, R, Budai, A, Hornegger, J, Jan, J, Gazarek, J, Kubena, T, Cernosek, P, Svoboda, O & Angelopoulou, E 2013, 'Retinal vessel segmentation by improved matched filtering: Evaluation on a new high-resolution fundus image database', IET Image Process., vol. 7, no. 4, pp. 373–383

68. Orlando JI, Fu, H, Breda, JB, van Keer, K, Bathula, DR, Diaz-Pinto, A, Fang, R, Heng, PA, Kim, J, Lee, J & Lee, J 2020, 'Refuge challenge: A unified framework for evaluating automated methods for glaucoma assessment from fundus photographs', Med. Image Anal., vol. 59, Art. Number: 101570

69. Pan, CW, Wong, TY, Chang, L, Lin, XY, Lavanya, R, Zheng, YF, Kok, YO, Wu, RY, Aung, T & Saw, SW 2011, 'Ocular biometry in an urban Indian population: The Singapore Indian eye study (SINDI)', Investigative Ophthalmol. Vis. Sci., vol. 52, no. 9, pp. 6636–664

70. Panda, R, Puhan, NB, Rao, A, Padhy, D & Panda, G 2018, 'Automated retinal nerve fiber layer defect detection using fundus imaging in glaucoma'. Comput Med Imaging Graph. vol. 66, pp. 56-65. doi: 10.1016/j.compmedimag.2018.02.006. Epub 2018 Mar 6. PMID: 29544118.

71. Parvin, H, Mirnabi Baboli, M & Rokny, HA 2015, 'Proposing a classifier ensemble framework based on classifier selection and decision tree', Eng. Appl. Artif. Intell., vol. 37, pp. 34–42. doi: 10.1016/j.engappai.2014.08.005

72. Punn, NS & Agarwal, S 2022, 'Modality specific U-Net variants for biomedical image segmentation: a survey'. Artif Intell Rev, https://doi.org/10.1007/s10462-022-10152-1

73. Raghavendra, U, Bhandary, Sulatha, Gudigar, Anjan, Acharya & Rajendra, U 2017, 'Novel expert system for glaucoma identification using non-parametric spatial envelope energy spectrum with fundus images'. Biocybernetics and Biomedical Engineering. 38. 10.1016/j.bbe.2017.11.002.

74. Raghavendra, U, Fujita, Hamido, Bhandary, Sulatha, Gudigar, Anjan, Tan, Jen Hong & Acharya U Rajendra 2018, 'Deep Convolution Neural Network for Accurate Diagnosis of Glaucoma Using Digital Fundus Images'. Information Sciences. vol. 441. 10.1016/j.ins.2018.01.051.

75. Raja, H, Akram, MU & Shaukat, A 2020, 'Extraction of Retinal Layers Through Convolution Neural Network (CNN) in an OCT Image for Glaucoma Diagnosis'. J Digit Imaging, vol. 33, pp. 1428–1442, https://doi.org/10.1007/s10278-020-00383-5

76. Raja, H, Hassan, T, Akram, MU and Werghi, N, 2020, 'Clinically verified hybrid deep learning system for retinal ganglion cells aware grading of glaucomatous progression', IEEE Transactions on Biomedical Engineering, 68(7), pp.2140-2151.

77. Sahu, S, Singh, AK, Ghrera, SP and Elhoseny, M, 2019, 'An approach for de-noising and contrast enhancement of retinal fundus image using CLAHE', Optics & Laser Technology, 110, pp.87-98.

78. Shahsuvaryan, Marianne & Ohanesian, Roger. 2005, Eye Diseases.

79. Shehryar, Tehmina, Akram, M, Khalid, Samina, Jameel & Amina 2017, 'Improved Automated Detection of Glaucoma from Fundus Image Using Hybrid Structural and Textural Features'. IET Image Processing. vol. 11. 10.1049/iet-ipr.2016.0812.

80. Shirzadfar, H & Amirzadeh, P 2019, 'A comprehensive study on eye issues and modern developments in visual rehabilitation for people with impaired vision'. Int J Biosen Bioelectron. vol. 5, no. 2, pp. 48-54. DOI: 10.15406/ijbsbe.2019.05.00153

81. Singh, A, Dutta, MK, ParthaSarathi, M, Uher, V & Burget, R 2015, 'Image processing based automatic diagnosis of glaucoma using wavelet features of segmented optic disc from fundus image'. Comput Methods Programs Biomed. vol. 124, pp. 108-20. doi: 10.1016/j.cmpb.2015.10.010. Epub 2015 Oct 23. PMID: 26574297.

82. Singh, Amitojdeep, Sengupta, Sourya, Lakshminarayanan & Vasudevan. 2019, 'Glaucoma diagnosis using transfer learning methods'. vol. 27. 10.1117/12.2529429.

83. Sivaswamy, J, Krishnadas, SR, Joshi, GD, Jain, M & Tabish, AUS 2014, 'Drishti-GS: Retinal image dataset for optic nerve head(ONH) segmentation', in Proc. IEEE 11th Int. Symp. Biomed. Imag. (ISBI), pp. 53–56

84. Sng, CC, Foo, LL, Cheng, CY, Allen, JC, He, M, Krishnaswamy, G, Nongpiur, ME, Friedman, DS, Wong, TY & Aung, T 2012, 'Determinants of anterior chamber depth: The Singapore Chinese eye study', Ophthalmology, vol. 119, no. 6, pp. 1143–1150

85. Soltani & Amira 2018, 'A new expert system based on fuzzy logic and image processing algorithms for early glaucoma diagnosis', Biomed. Signal Process. Control. vol. 40, pp. 366-377.

86. Song, W. Lai, TC & Su, YZ 2021, 'A Statistical Robust Glaucoma Detection Framework Combining Retinex, CNN, and DOE Using Fundus Images'. IEEE Access, vol. 9, pp. 103772-103783.

87. Staal, J, Abràmoff, MD, Niemeijer, M, Viergever, MA & Van Ginneken, B 2004, 'Ridge-based vessel segmentation in color images of the retina', IEEE Transactions Med. Imag., vol. 23, no. 4, pp. 501–509.

88. Tabassum, Munaza, Khan, Tariq, Arsalan, Muhammad, Naqvi, Syed, Ahmed, Mansoor, Ahmed, Hussain, Mirza & Jawad 2020, 'CDED-Net: Joint Segmentation of Optic Disc and Optic Cup for Glaucoma Screening'. IEEE Access. pp. 10.1109/ACCESS.2020.2998635.

89. Tékouabou, SCK, Chabbar, I, Toulni, H, Cherif, W & Silkan, 2022, 'Optimizing the early glaucoma detection from visual fields by combining preprocessing techniques and ensemble classifier with selection strategies'. Expert Systems with Applications, vol. 189, pp. 115975.

90. Tham, YC, Li, X, Wong, TY, Quigley, HA, Aung, T & Cheng, CY 2014, 'Global prevalence of glaucoma and projections of glaucoma burden through 2040: a systematic review and meta-analysis', Ophthalmology. vol. 121, pp. 2081–2090.

91. Tim McInerney & Demetri Terzopoulos 1996, Deformable models in medical image analysis: a survey, Medical Image Analysis, vol. 1, issue. 2, pp. 91-108, ISSN 1361-8415, https://doi.org/10.1016/S1361-8415(96)80007-7.

92. Veena, HN, Muruganandham, A & Kumaran, TS 2020, 'A Review on the optic disc and optic cup segmentation and classification approaches over retinal fundus images for detection of glaucoma'. SN Appl. Sci. vol. 2, pp. 1476. https://doi.org/10.1007/s42452-020-03221-z

93. Virbukaitė, Sandra & Bernataviciene, Jolita 2020, 'Deep Learning Methods for Glaucoma Identification Using Digital Fundus Images'. Baltic Journal of Modern Computing. 8. 10.22364/bjmc.2020.8.4.03.

94. Wang, S, Yin, Y, Cao, G, Wei, B, Zheng, Y & Yang, G 2015, 'Hierarchical retinal blood vessel segmentation based on feature and ensemble learning', Neurocomputing, vol. 149, pp. 708-717.

95. Weinreb, RN, Bowd, C, Moghimi, S, Tafreshi, A, Rausch, S & Zangwill LM 2019, Ophthalmic diagnostic imaging: Glaucoma. In: High resolution imaging in microscopy and ophthalmology'. Springer, Cham, pp. 107–134

96. Weiss, K, Khoshgoftaar, TM & Wang, D 2016, 'A survey of transfer learning'. J Big Data, vol. 3, no. 9. https://doi.org/10.1186/s40537-016-0043-6

97. Welfer, D, Scharcanski, J, Kitamura, CM, Dal Pizzol, MM, Ludwig, LW & Marinho, DR 2010, 'Segmentation of the optic disk in color eye fundus images using an adaptive morphological approach'. Comput Biol Med. vol. 40, no. 2, pp. 124-37. doi: 10.1016/j.compbiomed.2009. 11.009. Epub 2009 Dec 31. PMID: 20045104.

98. Yidong Chai, Hongyan Liu & Jie Xu 2018, 'Glaucoma diagnosis based on both hidden features and domain knowledge through deep learning models', Knowledge-Based Systems, vol. 161, pp. 147-156, https://doi.org/10.1016/j.knosys.2018.07.043

99. Yun, Jiang & Tan, Ning & Peng, TingTing 2019, 'Optic Disc and Cup Segmentation Based on Deep Convolutional Generative Adversarial Networks'. IEEE Access. pp. 1-1. 10.1109/ACCESS.2019.2917508.

100. Zhang, Z, Yin, FS, Liu, J, Wong, WK, Tan, NM, Lee, BH, Cheng, J, & Wong, TY 2010, 'ORIGA-light: An online retinal fundus image database for glaucoma analysis and research', in Proc. Annual International Conference., IEEE Eng. Med. Biol., pp. 3065 to 3068.

101. Zhao, Y, Rada, L, Chen, K, Harding, SP & Zheng, Y 2015, 'Automated vessel segmentation using infinite perimeter active contour model with hybrid region information with application to retinal images', IEEE transactions on medical imaging, vol. 34, no. 9, pp. 1797-1807.

www.ingramcontent.com/pod-product-compliance
Lightning Source LLC
LaVergne TN
LVHW082350200926
848481LV00012B/3522